DESIGN OF WATER QUALITY MONITORING SYSTEMS

DESIGN OF WATER QUALITY MONITORING SYSTEMS

Robert C. Ward
Colorado State University
Fort Collins, Colorado

Jim C. Loftis
Colorado State University
Fort Collins, Colorado

Graham B. McBride
Water Quality Centre
Department of Scientific and Industrial Research
Hamilton, New Zealand

VNR VAN NOSTRAND REINHOLD
New York

Library of Congress Catalog Card Number 90–46323
ISBN 0–442–00156–8

Manufactured in the United States of America

Published by Van Nostrand Reinhold
115 Fifth Avenue
New York, New York 10003

Chapman and Hall
2–6 Boundary Row
London, SE 1 8HN

Thomas Nelson Australia
102 Dodds Street
South Melbourne 3205
Victoria, Australia

Nelson Canada
1120 Birchmont Road
Scarborough, Ontario M1K 5G4, Canada

16 15 14 13 12 11 10 9 8 7 6 5 4 3 2 1

Library of Congress Cataloging-in-Publication Data

Ward, Robert C.
Design of water quality monitoring systems / Robert C. Ward, Jim C. Loftis, Graham B. McBride.
p. cm.
Includes bibliographical references and index.
ISBN 0–442–00156–8
1. Water quality—Measurement. I. Loftis, Jim C. II. McBride, Graham B., 1948– . III. Title.
TD367.W36 1990
628.1′61—dc20 90–46323
CIP

PREFACE

As human impact on the environment becomes increasingly life threatening, there is an ever louder call for better management of our planet's resources. Professionals, involved in management to improve our stewardship over the earth, increasingly face the difficult task of integrating law, science, political action, and public reaction into a logical framework. For this integration to occur successfully, management must have an accurate and adequate supply of information about the behavior of the environment under constantly changing human pressures and natural forces.

In the field of water quality management, there has been a steady evolution in procedures for designing systems to obtain information on the behavior of water quality in the environment. This evolution began with site specific efforts to obtain information to solve specific problems and has expanded to now include efforts by large local, national and worldwide organizations to assess water quality behavior within their jurisdictions. These efforts to better describe the behavior of water quality over large expanses of space and time have faced a number of problems, often related to a lack of both scientific understanding and strong organizational underpinnings. Realization of this fact has led to considerable research on the application of science and better organizational skills needed for the design of water quality monitoring systems.

The knowledge being developed is published in separate research subject areas. The means for graphically and statistically analyzing water quality data are being improved and better documented in journal articles and statistical textbooks. More detailed and comprehensive laboratory analysis protocols are being developed and published as handbooks and guidance documents. Computers are increasingly being incorporated into the operation of water quality monitoring systems, but documentation on this subject is scattered.

A person assigned the task of designing or operating a water quality monitoring system sees a lot of information on specific aspects of monitoring, but it is not clear how all the existing design and operational information is tied together. Neither is it clear how monitoring tasks are to be connected such that the end product meets the information needs of management.

If a book were to be written to tell monitoring system designers how to approach the organization and execution of the design process, what would it cover? How would it be organized? Would it cover all the "hard" science of design and/or the "soft" aspects of information use for management decision making? These, and many more, questions faced the authors of this book.

In the end, we chose to focus on the process of design rather than on the details of each step. We felt that the details are, for most components of the system, being adequately addressed (individually) in the literature. What was needed most was a framework explaining how a monitoring system designer could tie all the parts together. This book represents our efforts in that regard.

We chose to rely heavily on references to existing literature. Some literature, viewed as critical to explaining the total systems approach, is reviewed in the book, while other literature that has reached a fairly well accepted and standardized status, is simply referenced. Because of the large amount of knowledge required to design a total water quality monitoring system, the readers of this book will find themselves needing to refer to other literature frequently. It simply is not possible to include everything in one book. Hopefully, this book will help designers to access the relevant literature quickly and efficiently.

While we strongly emphasize the need for a total systems approach to design, this book focuses on the "information end" of the water quality monitoring system—the data analysis, reporting, and use of the resulting information within management decision making. The recommended approach calls for clear definition of information goals along with specification of data analysis methods to be employed to reach those goals.

To many readers the word "monitoring" may not, at first, imply the depth of effort being suggested in this book with its emphasis on acquiring water quality information as opposed to data. It is probably more definitive to talk about the design of "water quality information systems" when the end product of the measurement process is knowledge of the behavior of water quality in the environment instead of data that is simply compared to a standard or perhaps filed away and forgotten. However, we use the term "monitoring" in the broader sense to describe the overall act of obtaining "information" about water quality.

In presenting design approaches, it is easy to get lost in statistical details and not spend enough time on the more difficult task of explaining the information context within which statistics are to be used. While writing the book, we often felt the tug to simply explain statistical calculations and to only touch upon their relationships to information objectives. We have resisted this temptation and have avoided repeating details which can be found in most basic statistics books. We have included many references which are relevant to water quality monitoring, but we rely most heavily upon Richard Gilbert's 1987 text entitled *Statistical Methods for Environmental Pollution Monitoring*.

The book's organization reflects our attempt to address the "information" end of the total monitoring system. After Chapter 1 reviews the evolution of water quality monitoring as an information system, the subsequent chapters first define a water quality monitoring system and then describe how one can approach the design of such a system. Chapter 2 defines water quality monitoring as a system following the flow of information through six major components. The focus on insuring that useful information is obtained from the monitoring effort leads to a description in Chapter 3 of the importance of statistics in the total design effort. The importance of the role of statistics in monitoring is basic to the philosophy that underlies the entire book. Chapter 4 outlines an overall approach to designing a total water quality information system and sets the stage for the details presented in Chapters 5, 6, and 7. The design culminates in a final design report, described in Chapter 8, which documents the entire effort and insures that the design process will more than return its considerable investment.

Chapter 9 then presents several case studies which describe efforts of various organizations and individuals to design water quality monitoring systems using many of the concepts advocated in this book. The case studies point out very clearly that the design of water quality information systems still is as much an art as it is a science and still requires considerable ingenuity on the part of the designer. Chapter 10 closes the book with a brief summary and final system design check list.

The authors wish to acknowledge the support of their employers in preparation of this book. Colorado State University provided a sabbatical to Robert Ward in 1983–1984 during which time he worked with Graham McBride at the Water Quality Centre in Hamilton, New Zealand, in initiating the design of a water quality monitoring program for the country. Upon returning to Colorado State, Robert, in conjunction with Jim Loftis, developed an experimental course on the topic of design of water quality monitoring systems. The course was taught by Robert and Jim in 1985 and 1987 and by Graham in 1989. The notes developed for these offerings of the course have led to the writing of this book. Graham spent the last half of 1989 in Fort Collins, Colorado, on leave from the Water Quality Centre; and it was during this time period that the vast majority of the book was developed.

We also acknowledge support from monitoring system managers working in the monitoring ''trenches'' as their input over the years has greatly influenced the contents of this book. We also want to thank our students—who have labored with a loose set of notes which were under constant revision—for their patience and helpful comments. Finally, we have benefited greatly from outside reviewers, but any errors or omissions that remain are entirely our responsibility.

CONTENTS

ACKNOWLEDGMENTS

The authors would like to thank their employers, Colorado State University and the Water Quality Centre, Department of Scientific and Industrial Research, for their support of the preparation of this manuscript and of the research and teaching which led to it.

We would also like to thank our spouses, Brenda Ward, Judy Billica, and Robyn McBride for their support and patient understanding of our preoccupation with water quality monitoring.

DESIGN OF WATER QUALITY MONITORING SYSTEMS

1

INTRODUCTION

Society, as it strives to achieve ever increasing standards of living, has an insatiable demand for clean water, both for its own consumption and enjoyment in the environment. Too often, creation of a higher standard of living carries with it an environmental "price" in the form of water pollution. To reduce this "price" and, at the same time, spread it equally among all water users, society passes laws to manage water quality. The exact form and nature of these laws has evolved over the years. Each law contains a number of management functions and powers with which water quality is to be controlled. For example, a water quality management agency may be given the function to establish water quality criteria which are eventually incorporated into water quality standards. Since the criteria and standards are called for in the law, they have a legal basis for use in water quality management.

When a law establishes a water quality management agency and gives it the responsibility to manage water quality for the benefit of society, there is often mentioned the need to measure, or monitor, water quality. Too often, however, little mention is made of how the information on the quality of the water is to be used in making decisions regarding implementation of the management functions and powers provided in the law. It is left to the professionals hired by the government to implement the law to decide how the results of monitoring are to be used in management decision making.

The science that serves as the basis of water quality monitoring is evolving rapidly. This science is necessarily broad and complex. It has to cover the nature of environmental decision making, aquatic ecology, the statistics of analyzing data, the chemistry of water, the toxicity of chemicals to biological organisms, data management hardware and software, large scale hydrology, and many other areas of science. Invariably the people asked by society to implement water quality management laws and their resulting water quality monitoring systems are not expert in all areas necessary for the successful design, implementation, and operation of a total water quality information system. There are few courses available on the design of water quality monitoring systems!

The purpose of this book is to bring together current knowledge on the design of water quality monitoring systems and present it in a form which is useful to professionals charged with designing, operating, and managing such systems. The book emphasizes the need to focus the design on explicit information goals. The book does not attempt to spell out all the details of sampling, laboratory analysis, and quality control/ quality assurance—topics that have been standardized in many guidebooks. Instead, a framework is developed within which these topics become major elements in a total information system.

The book does not focus on water in one portion of the hydrologic cycle, e.g., ground water. Instead, it focuses on generating information on water quality that can be obtained from a data record, regardless of where the data come from. For specifics of actually obtaining a sample from water in a particular part of the hydrologic cycle, the authors encourage use of readily available standard procedures published by government agencies and professional societies.

This work is intended as a companion reference to that of Gilbert (1987), which gives extensive discussion of statistical methods for environmental pollution monitoring. The authors have, therefore, attempted to be consistent with his conventions, nomenclature, and recommended approaches. However, experience in teaching courses on this topic has shown that many readers are likely to lack confidence in their understanding of basic statistical theory. This is due partly to a lack of formal training in the topic and partly to the profusion of different conventions for nomenclature and for reading of statistical tables. For example, one text will use upper case letters for sample statistics, another will use lower case: one will report right-tail areas in its tables, another will report both tails. The resulting confusion is enough to deter many from the use of statistics. The authors simply recommend that, when necessary, the readers consult a good basic text, and be aware that conventions differ from one text to another.

The authors have included a special chapter (chapter 3) which addresses two statistical concepts critical to the general area of statistical inference: estimation and testing. They find considerable confusion in the literature on the meaning and application of hypothesis tests, and, in particular, on the meaning of the term "significant differences." An understanding of these concepts is essential for stating explicit information objectives.

In the same chapter, the often-neglected requirement for tests to have good statistical power is discussed (Millard 1987, Cohen 1988).

While Gilbert (1987) presents the mechanics of the statistics utilized in the discussion of water quality information system design, Sanders et al. (1983) describe the means by which a monitoring system designer can quantitatively approach the design of an actual monitoring network. In particular, Sanders et al. (1983) present methods to determine where to sample, computations to determine sampling frequencies, and considerations for the selection of water quality variables to measure. This earlier work does not, however, develop the concept of a water quality information system of which the network is only a part—thus the need for this book.

1.1 EVOLVING NATURE OF WATER QUALITY MONITORING

A quick review of the water quality laws in many countries will reveal the evolving nature of water quality management. For this discussion, the authors focus on experience in the United States. Most early efforts to manage water quality in the United States focused on preventing nuisances. Keeping waterways free of floating debris and foul odors for the purposes of navigation were the basis of early laws. At the beginning of the 20th century, relationships between water quality and public health began to influence legislation. The focus of the laws began to shift toward pollution control. This focus on pollution control, as opposed to a broader approach to management in general, carried through World War II, even though many early arguments for the basis of modern water quality management began during the 1930s. It was not until the 1940s and 1950s that Federal laws (PL 80-845 and PL 84-660) on water pollution control were actually passed in the United States. Many states had already developed their own water pollution control strategies, and there were considerable differences in approach.

Velz (1950) and Haney and Schmidt (1958) represent two early efforts to develop systematic approaches to water quality monitoring. Velz (1950) notes that monitoring for effective evaluation of water quality conditions involves three operations: (1) Collection of data; (2) Analysis of data to determine specific self purification characteristics; and (3) Projection of estimates for other conditions. He concludes that a statistical approach to monitoring affords a maximum of information with a minimum of cost in time and funds. Clark (1950), in a discussion of Velz's paper, makes the following statement:

> "In considering the desirability of using a statistical approach to stream analysis, it must be kept in mind that a stream presents a maze of variables—physical, chemical, and biological—the analysis of which demands the application of the laws of probability. Under such conditions, a single observation is virtually worthless; the development of a true picture and the projection of a reliable prediction require the careful and accurate application of statistical laws."

Until the 1956 Federal Water Pollution Control Act (PL 84-660) in the United States, the Federal laws had not specifically required water quality monitoring. Of course monitoring had been performed during the course of implementing the laws, but it was usually tied to a site specific analysis of a problem and was limited in the number of physical and chemical characteristics measured.

With passage of the 1965 Federal Water Quality Act (PL 89-234), a common approach to water quality management was forced upon all state water quality management efforts in the United States. As with earlier laws, the 1965 Act focused on surface water quality. The Act also spelled out the need for all state water quality agencies to begin monitoring. Beyond the simple requirement to monitor, the law did not spell out why the monitoring was needed or how it was to be performed. Most states reacted by initiating some form of regular water quality monitoring to meet the requirements of the Federal law and, thus, qualify for the Federal monies to support planning and building of wastewater treatment plants. Also, if the states did not comply with the Federal requirements, the Federal government had the responsibility to assume control of water quality management within the state.

Without much guidance or a strong body of knowledge to draw upon, the designers of these early water quality monitoring efforts were often learning by doing. The knowledge they gained, however, set the stage for identifying the weak areas of monitoring system design and began the development of design procedures that the authors are now attempting to present in a coordinated fashion. It was during this time that the need for a computerized data base for storing water quality data led to creation of STORET (a federally supported water quality data *sto*rage and *ret*rieval system that continues to exist and evolve today).

As states implemented provisions of the 1965 Federal Water Quality

Act, it soon became obvious that the monitoring programs were not providing the information needed by the new management strategy. Studies to examine the design of water quality monitoring systems (or surveillance as it was commonly referred to then) were initiated by forerunners of the U.S. Environmental Protection Agency (EPA) in the late 1960s and early 1970s. Kittrell (1969), Morgan et al. (1970), Beckers et al. (1972), Ward (1973), and Beckers and Chamberlain (1974) describe these early efforts to develop better water quality monitoring systems. Pomeroy and Orlob (1967) describe an attempt by the State of California to identify a role and design for water quality monitoring in its management system. The proceedings of a Wisconsin symposium (Kerrigan 1970) addressed many of the water quality monitoring/management issues of the late sixties and early seventies, in particular the benefits offered by advanced data collection, analysis, and display systems toward the enhancement of water quality management efforts.

Much of the monitoring literature at this time was in the same vein as past monitoring efforts—that is to determine whether water quality is sufficient for a set of designated uses. Samples were collected, checked against criteria for an established use, and determined to be acceptable or non-acceptable. Questions were being raised about the use of such information within water quality management strategies (Kerrigan 1970), but the strong focus on sample collection and laboratory analysis procedures meant that most information questions were not answered.

While the 1965 Act led to increased water quality monitoring in the United States, the 1972 Amendments to the Federal Water Pollution Control Act (PL 92-500) led to a decreased interest in the results of stream monitoring. The 1972 Act shifted the focus of water quality management, for a time, from control of in-stream water quality conditions to control of discharges to streams via the newly created discharge permit system. Implementation of the permit system consumed considerable management effort and reduced the reliance of management decision making and enforcement on in-stream data and information. Instead data from end-of-pipe monitoring became the focus. In-stream monitoring did not disappear, but questions about its role in management began to surface. Section 305(b) of the 1972 Act required states and the U.S. EPA to prepare annual reports on the status of water quality in the United States. Ongoing, in-stream monitoring was often viewed as necessary for this reporting requirement, although it was not until the 1980s that much effort was put into preparing meaningful 305(b) reports. Schnider and Shapiro (1976) is a manual designed to evaluate the efficiency of surface water quality monitoring networks to document compliance with ambient water quality standards. It does not address the reporting requirements of Section 305(b) but rather focuses on sampling and laboratory analysis.

During the mid-seventies, the quality of waters in other parts of the hydrologic cycle began to attract the attention of society, and new laws focused on drinking water quality, ground water quality, and acid precipitation. As the problems associated with these waters often involved trace amounts of toxic chemicals, the list of conventional chemicals that had been the focus of water quality monitoring in surface water was now augmented by a large number of additional water quality measurements. While surface water quality monitoring was still being conducted, these new monitoring requirements often forced reductions in the number of sampling sites or samples taken in the streams. The competition for monitoring resources began to bring to light many questions regarding the role of monitoring within a water quality management agency.

What understanding of water quality in the hydrologic cycle is necessary to write a discharge permit? What understanding of historical water quality conditions is necessary to set water quality standards on a water quality limited stream segment (a stream segment whose use is limited by preexisting water quality conditions)? What understanding of water quality trends is necessary to prepare an informative assessment of water quality conditions over time (per 305(b) requirements)? Work on the design of water quality monitoring systems focused around these types of questions during the 1980s.

As the need to account for the stochastic (statistical or uncertain) nature of water quality began to be recognized within the water quality management community, studies of water quality monitoring system design began to include a more statistical focus. Sanders (1974), Lettenmaier (1975), Sherwani and Moreau (1975), and Loftis (1978) are examples of early efforts to introduce stochastic considerations into the design of water quality monitoring systems. The World Health Organization (1978) is a guide for the collection and interpretation of water quality data that addresses the statistical nature of data interpretation. Montgomery and Hart (1974), and the discussion of their paper (attached to the end of the paper), present a review of the evolving use of statistics in water quality data analysis and indicate the need for further research on the topic.

The effort to better quantify the information needed to manage water quality, and, therefore, establish more scientific design criteria for monitoring system design, has evolved from a number of observations and studies in the late 1970s and 1980s. Hines, Rickert, and McKenzie (1977), in studying a specific water quality problem, noted:

> "In many basins it has proven difficult to use existing river-quality data for analysis of temporal and spatial trends. . . . Major reasons for this problem are the arbitrary nature of sampling programs that generate the quality data and a general failure to account for the background variability in quality resulting from hydrologic phenomena."

Also, in the same year, the National Academy of Sciences (1977) noted a general lack of the use of statistics in designing what is, in reality, a statistical sampling process.

The Council on Environmental Quality (1980) called for more coordination in water quality monitoring, implying that there are too many data and not enough information. The General Accounting Office (1981), in noting the lack of information from routine, fixed-station monitoring (regular sampling at set sites with no specified time of termination), called for its elimination in favor of short-term, special surveys which would be repeated in each basin every five years. Loftis and Ward (1982) and van Belle and Hughes (1983) both argued that routine monitoring provides information on the behavior of water quality that cannot be obtained via special surveys only. In particular, routine monitoring is better able to assess background variability, seasonality, and long term trends. Both papers note past problems with the design of routine monitoring programs including lack of clear objectives, failure to use statistics, and the ad hoc nature of the designs. However, they argued for better designs, not the elimination of an important component of a total information system needed to manage water quality.

The Committee on Science and Technology (1983) focused national attention on the need for better water quality monitoring system design. In a second report on the subject, the General Accounting Office (1986) critiqued five national water quality monitoring efforts established to assess water quality nationally. In this report, the General Accounting Office reversed itself from the 1981 report and called for better trend assessment monitoring. The report makes the point that information on water quality conditions over time is needed for decision making. The National Research Council (1987) reviewed the issues surrounding water quality monitoring and strongly encouraged the development of better information for water quality management decision making.

The U.S. EPA (1987a) published a "framework" for change in surface water monitoring. The purpose of the report was to initiate an assessment of surface water monitoring with the goal of improving the design of such systems and, ultimately, of improving the information generated for water quality management decision making. This report is the latest in a series of reports that the U.S. EPA has issued to help states establish better water quality monitoring systems. The earlier reports are U.S. EPA (1975, 1977, and 1985b).

These critiques and efforts to improve water quality monitoring (information) systems often identify the lack of quantitative, systematic design procedures. The state-of-the-art in water quality monitoring system design has evolved to the point where it is possible to describe approaches to design. This is not to say that more evolution is not needed. In fact, as the design of monitoring systems becomes more quantitative, areas of needed research and development become more obvious. Bloem and Glicker (1989) make this point very clearly as they describe the evolution of the Portland Water Bureau's water quality monitoring program associated with its public water supply. They note that the development of a water quality information system is iterative. As the water quality behavior is better understood, new questions arise requiring modification in the design of the system. These design modifications often require research into the nature of the question and the type and form of the information that will provide the needed answers.

Efforts to improve the design of water quality monitoring systems have gained momentum in recent years. Among the more notable examples are the following: Hirsch et al. (1988) and Payne and Ford (1988) are descriptions of efforts, by the U.S. Geological Survey and the U.S. EPA, respectively, to design national water quality monitoring efforts with specific information goals. Both reports, which have the word "concept" in the title, describe the logic and science to be used in designing the respective monitoring efforts. The proceedings of the International Symposium on the Design of Water Quality Information Systems (Ward, Loftis, and McBride 1989), contain 44 papers discussing aspects of designing total water quality monitoring systems. Smith et al. (1989) present the documented design of a national water quality network for New Zealand. In a series of three issues in 1985, the *Water Quality Bulletin* (Vol. 10, Numbers 2–4), describes water quality monitoring efforts in many countries around the world, as well as the GEMS (Global Environmental Monitoring System) program.

Thus, the evolution of water quality monitoring system design has been underway for some time and appears to be gaining momentum as water quality management is becoming increasingly sophisticated in its need for and use of information about the behavior of water quality conditions in the environment. Hopefully, this book represents a further step in the evolution of water quality monitoring system design.

1.2 CURRENT DEFINITIONS AND TYPES OF MONITORING

Whenever a system is to be designed, it is critical that the system first be defined. While this may sound obvious, it has not often been the case in water quality monitoring. Past design practices have generally been on an ad hoc basis with no design documentation prepared. The design approach too often involved several individuals deciding where to collect samples (based on ease of access), what to measure on each sample (based on laboratory capability), and how often to collect the samples (based on the budget available and scheduling of collection routes around the agency's other management tasks). The reasons for the decisions made were seldom documented. Whenever a crisis arose in any portion of the management agency, monitoring activities were the first to be curtailed to obtain the staff time to address the problem. Whenever a budget problem arose, monitoring was generally the first activity to be considered for cuts. *Without a clear and documented reason for existing,* water quality monitoring was an easy target for cuts in budget and personnel. It was almost as if monitoring efforts were storing money and personnel for later use, as needed, by the agency.

This lack of a clear purpose and documentation of monitoring led to considerable data being collected without a means of converting them into information and without a defined use within the decision making activities of the agency. The authors previously noted the situation as one of "data rich but information poor" (Ward, Loftis, and McBride 1986). To begin to address the problem, the concept of a "water quality information system" was proposed by Ward (1979) and Ward (1986). These concepts have been reinforced by much of the "information revolution" thinking that was initiated by Naisbitt (1982). Many of his ideas can be applied directly to the information problems associated with water quality monitoring. For example, his quote:

> ". . . uncontrolled and unorganized information is no longer a resource in an information society. Instead it becomes the enemy of

the information worker . . . information technology brings order . . . and therefore gives value to data that would otherwise be useless.''

relates very well to the situation in water quality monitoring today.

Stednick and Roig (1989) note that ''almost all types of water quality sampling or measurement are termed water quality monitoring.'' Thus, if one sample is taken at a randomly selected point to check compliance with a standard, such an effort may be referred to as ''water quality monitoring.'' On the other hand, the specification of 600 sites across the country to be measured monthly may also be referred to as ''water quality monitoring.'' Under such a situation, it becomes very important that the designer of a water quality monitoring system first define exactly what he/she is designing!

There are many forms of water quality monitoring. They can be categorized in several ways including: 1) length of projected life of the monitoring system; 2) types of measurements to be made; 3) location of the water in the hydrologic cycle; and 4) type of water quality management tool to be supported.

If a water quality monitoring effort consists of fixed sites where samples are being taken and there is no projected ending point for the sampling (similar to the flow measurement networks operated by many hydrological agencies around the world), then the monitoring effort is often referred to as an ambient, fixed-station, or long-term monitoring program. Generally, these water quality monitoring efforts attempt to describe water quality conditions over large areas and over long periods of time. Many of the early water quality monitoring efforts were of this type as they followed the general scheme used in hydrology to measure flows.

If there is a set time frame for a water quality monitoring effort, it is often referred to as a special study, intensive survey, or short-term monitoring effort. This type of monitoring generally has a specific information purpose related to a particular water quality problem that needs further definition in order to develop a solution. Or it may be associated with the need to model a segment of a river to perform a wasteload allocation as part of issuing a permit. Unfortunately, fixed-station monitoring often does not have a well defined reporting mechanism in place. Special study monitoring, on the other hand, more often requires that a report be prepared following completion of the monitoring efforts.

The exact distinction between these two types of monitoring, has been blurred with recent proposals to study watershed units with intensive monitoring efforts for a 2 to 5 year time period, followed by periods when the monitoring will be less intense (Hirsch et al. 1988; Cohen, Alley, and Wilber 1988; and Payne and Ford 1988). To a certain extent this is a blend of fixed-station monitoring with special studies which appears to take advantage of the strengths of each.

If types of measurements are used to categorize monitoring, one has the obvious split between physical/chemical and biological monitoring. There are many subsets of each which further define a type of monitoring. For example, toxics monitoring and nutrient monitoring refer to efforts that focus on specific chemicals in the water. Within the biological monitoring category, there are such terms as fisheries survey, bioaccumulation surveys, bioassessments, and ecosystem monitoring.

If location of water in the hydrologic cycle is used to categorize monitoring systems, then such terms as ground water monitoring, lake monitoring, acid rain monitoring, surface water monitoring, effluent monitoring, estuary monitoring, and marine monitoring are used.

Monitoring may also be categorized by the utilization of the monitoring results within management. The resulting classifications include compliance monitoring, enforcement monitoring, trend monitoring, and background monitoring.

From this discussion one may begin to see why design of water quality monitoring systems is not straightforward. It is very important that the object of the design exercise be carefully defined. For purposes of this book, the categorization of monitoring systems based on design life (category 1 above) will be used to organize the discussion.

For example, the following three reports describe what will be referred to as long term monitoring. Smith and McBride (1989) describe the design of a national stream water quality monitoring system in New Zealand; Bell (1989) describes the evolution of a ground water quality monitoring program for an IBM semi-conductor manufacturing site; and Bloem and Glicker (1989) and Tracy and McCammon (1989) describe a similar evolution of a public utility's monitoring program for its raw water supply. (The New Zealand and IBM monitoring systems are described in Chapter 9.) These monitoring system design efforts all tend to focus on the long term need for water quality information and, as such, do not have defined time limits.

At the other end of the spectrum, the following three reports describe special studies. A special study design used to investigate a reported spill was reported by Lawrence et al. (1989). Ponce et al. (1982) describe a design approach for studies of forestry operations on water quality in national forests. Rickert and Hines (1975) describe a framework for assessing river water quality developed as part of the U.S. Geological Survey's intensive assessment of the Willamette River basin in Oregon in the mid-seventies.

Other classifications of monitoring systems are just as valid, and there is no attempt on our part to exclude any water quality measurements (including biological monitoring measurements) from the design exercise. The design methodologies can be applied to water in any portion of the hydrologic cycle, and the resulting information can be used to support a number of management activities. Using design life, or time, as the means of categorizing monitoring does, however, require that the resulting types of monitoring be related to the various purposes of monitoring described above.

For example, a special survey can include physical/chemical measurements as well as biological measurements. A special survey can be used to obtain enforcement information. A special survey can be focused on ground water quality evaluation. Likewise, a fixed-station monitoring effort can include a range of measurements, provide information for setting standards, or be focused on acid deposition impacts. The design of the monitoring system will reflect its specific purpose and focus. Too often designs have not been documented, and the various terms noted above have been used to help describe a monitoring system. Unfortunately, adjectives in front of the word ''monitoring'' are not sufficient to define its purpose and focus.

1.3 ROLE OF MONITORING IN WATER QUALITY MANAGEMENT

From the law arise various water quality management strategies—each with its own way of using the results of water quality monitoring. The spectrum ranges from a ''crisis management'' approach to the ''no-pollutant-discharge'' approach. Crisis management generally uses monitoring on a special study basis and is often utilized when there are limited resources available. It also represents some of the earlier forms of water quality management from which more sophisticated approaches have subsequently evolved. The no-pollutant-discharge approach, on the other hand, tends to focus its monitoring efforts on discharges, not on the receiving water.

In practice, most water quality management agencies operate somewhere between the above two extremes. The exact nature of information needed on water quality is a function of administrative interpretation of the law. As there may be numerous implications in a given law regarding the need to understand the behavior of water quality, it is up to the staff of each agency to decide on what they want to emphasize in monitoring. Needs for understanding water quality behavior may be implied in *legal goals, functions and powers of management* provided to reach the goals, or they may be *directly stated* as monitoring requirements. To more fully appreciate these implications, first consider the legal goal of "fishable and swimmable" (PL 92-500). How does an agency, particularly a state-wide or nation-wide agency, obtain information on water quality that permits it to state whether the goal is met? Is there a clear relationship between the goal and the data obtained to measure success in its accomplishment? Does the agency even need to measure success in meeting legal goals?

Next consider that most water quality management laws provide a number of management functions and powers (management "tools") to be utilized in managing the quality of water. Each of these tools requires information on the quality of the water if it is to be implemented fairly and effectively. The exact information needs of each tool are not usually spelled out in the law but rather are defined by the staff charged with implementation. These water quality management professionals must often answer difficult questions, such as the following. How does an agency identify exactly what information it needs to properly implement its management tools? What information is needed to state quantitatively that stream standards are met? What water quality information is needed to write a discharge permit? What water quality information is needed to issue a loan for construction of a wastewater treatment plant?

Workable answers to these questions can be derived from the management strategy chosen to implement the law. To design water quality monitoring systems that support water quality management, this strategy must be identified and its information needs defined. The connection between the *data collected* by monitoring and its *use* within a management agency is an important element in the success of any water quality monitoring system design.

Finally, beyond the above implications for monitoring within a water quality management agency, there are directly stated monitoring requirements in many water quality management laws. For example, section 305(b) of the Federal Clean Water Act requires that states prepare a report which describes the water quality of all navigable waters in the state. This rather straightforward requirement for ambient water quality monitoring is often the justification for fixed-station monitoring operated by states. In this role, assessment monitoring is used to determine whether legal goals are being met. There is, however, not a clear design connection between most ambient monitoring systems and goal assessment. For this reason, assessing goal achievement with ambient monitoring data is difficult. The need for current research and development in this area is evidenced by the continuing evolution of 305(b) report guidance by U.S. EPA and the current redesign of the NASQAN (National Stream Quality Accounting Network) program by the U.S. Geological Survey.

Thus, the role of monitoring in management, while being weakly guided by the law setting up the management agency, is very much open to interpretation by the personnel implementing the law. Defining the role the staff has chosen to employ is the first step in designing an effective water quality monitoring (information) system that supports management decision making in a quantitative manner. This topic is revisited in Chapter 5 where the formulation of information objectives is discussed in more detail.

1.4 THE NEED TO DESIGN MONITORING SYSTEMS

Without carefully connecting the acquisition of data with the production and use of information from the data, there is a high probability that data collection becomes an end in itself. Whenever the conversion of data to information is contemplated, one might reflect upon exactly what the data represent. Each sample that results in a data point after laboratory analysis represents only a very small portion of the total water "population." If it is desired to make inferences about the total water quality situation in a large body of water (either over time or space), then some means of abstracting general characteristics from the data must be utilized. Simply presenting tables of data (still a widespread practice) does not convey general characteristics. Graphical representations are often used to show water quality conditions, but these may not permit quantification of the significance of trends, probabilities of stream standard violations, or the precision of estimates of average conditions. To obtain this quantification, statistical methods must be used. Statistics, therefore, becomes the vehicle for connecting and quantifying management information needs with the results of monitoring data.

As Zar (1984) notes, a knowledge of basic statistical principles and procedures is important even before data collection begins. There has to be a clear definition of the "population" to be sampled. Gilbert

(1987) notes that "unless the population is defined and related to the study objectives and field sampling procedures, the collected data may contain very little useful information for the purpose at hand." Too often in the past, water quality data were collected until such time that someone asked what the data "said." The entire data record was then taken to a statistician who was then requested to answer this question. Obviously, an effective water quality information system does not work this way, never has, and never will! If a management agency wants answers to water quality questions, then the questions will have to be quantified and a monitoring system (including its statistical methods) designed before the data are collected.

The realization that water quality is often highly variable must be recognized and accounted for in the statistical design. It simply is not possible to measure all conceivable water quality events; and, since exact laws of cause and effect are generally unknown, unexplained variability will exist in the data (Steel and Torrie, 1960). The water quality manager then must extrapolate or make inferences from particular samples to more general conclusions. The field of statistics provides us with an organized approach to quantify the unavoidable uncertainties about the inferences drawn from water quality data. In this light, Snedecor and Cochran (1980) define statistics as a field that deals with techniques for collecting, analyzing, and drawing conclusions from data. This is exactly what is desired in any water quality monitoring effort. The only way to ensure that accurate conclusions will be developed from water quality data is to bring statistics into the initial design of the monitoring system.

Steel and Torrie (1960) relate that statistics implies, for most students, a new way of thinking—thinking in terms of uncertainties and probabilities. Many professionals managing water quality monitoring systems have had little or no statistical training before they find themselves designing water quality monitoring systems. This situation develops as a result of monitoring not being viewed as a system needing "design." If design is viewed as only deciding where, what, and when of sample collection, then there is not a great need for statistics. However, if there is ultimately to be information obtained from the data, then a monitoring system has to be designed in its entirety.

How much time and money should be spent designing a monitoring system? If a thorough design is to be developed (per the procedures described in this book), it would not be unrealistic to use a figure that represents 10% of the projected 20 year operating cost of the monitoring system (for a long-term monitoring effort). One would never consider building a bridge or high-rise structure without hiring a design expert who would spend 10% to 15% of the capital construction cost on the design exercise. Why are thousands and even millions spent on water quality monitoring without designing the information system? "Blueprints" are also needed to describe how to build a bridge or building, how to collect samples, analyze them in the lab, store and retrieve the data, analyze the data, report the results, and then use the information in making management decisions.

Too many monitoring systems, in the past, have consumed large sums of money only to produce little or no information on water quality. Thus, water quality monitoring systems must be designed, and there should be a design document that describes the operation of every part of the system. Only in this way can the people who are to pay for the monitoring be assured that they will receive their money's worth.

2

DEFINITION OF THE MONITORING SYSTEM

To design a water quality monitoring system, the system must first be identified and defined in such a way that all its various components can be related to an overall information goal. With all the various types and purposes of monitoring, noted in Chapter 1, it is easy to see why there is no consensus as to what a water quality monitoring system should look like. Past definitions of water quality monitoring focused on the actual means for collecting data and not on the information purpose for monitoring. To define a complete monitoring system, however, there should be components that deal with both data collection and information generation.

In keeping with the focus of this book, the definition of a monitoring system will be based on the flow of "information" through a series of monitoring system components. The flow of information can be viewed as beginning with the actual interface between the water and monitoring system personnel—the collection of the sample. For purposes of this discussion, no specification of measurements is being made. Measurements can be physical, chemical, or biological in nature. Once a sample is collected, it can be analyzed in the field or taken back to a laboratory. Thus, if the sample, or the information it contains is followed, a water quality monitoring (or information) system can be defined.

Continuing the above process, one sees the sample going to the laboratory where it is analyzed. The laboratory results (numbers) now represent the information being followed. The water quality information in the sample has, via laboratory analysis, been converted into data. Once the data have been confirmed, the sample collected in the environment is discarded. These laboratory results, or data, are now stored in some form of data storage and retrieval system. Such a component within an information system permits the data to be organized and readily available for further analysis and conversion to additional information.

At some point, sufficient data are available to support data analysis, either via graphical presentation, statistics, modeling, or some combination of these. The choice of data analysis methodology depends upon the information sought. Ideally, the data analysis methods have been identified prior to sampling so that the data are collected in direct support of the data analysis methodology.

After the data are analyzed, the results of the data analysis must be reported to the information user. Reporting can take many forms, depending upon the information need, timeliness sought, and the management style of the information users.

Generation of reports cannot be viewed as the final component in the water quality monitoring (information) system. Unless the information is used, the system is incomplete. Thus, "information utilization" becomes the final component in the water quality information system. Defining how water quality information will be used within a water quality management program is an essential component of the design. Without quantification, the information may or may not be used within management. If it is not to be used, why spend money to collect it?

The water quality monitoring system, following the flow of information, can then be defined in a summary fashion as in Figure 2.1.

Thus, the purpose of a monitoring system is to develop an understanding of the water quality conditions that exist in the water body of interest and thereby facilitate sound management decisions. The monitoring system is the only connection between the water in the environment and decision makers.

Defining the water quality monitoring system (see Figure 2.1) permits the organization of data and information that Naisbitt (1982) indicated was necessary to give value to data and information. If the system definition is used to guide its design and operation, the chance that information loses its connection with its original information objective is greatly reduced.

2.1 DATA COLLECTION VERSUS INFORMATION GENERATION

With the monitoring system defined as in Figure 2.1, the first three components can be viewed as the data generation portion of the total monitoring system. (Traditionally, the first three components have been viewed as the total system because the acquisition of data was viewed

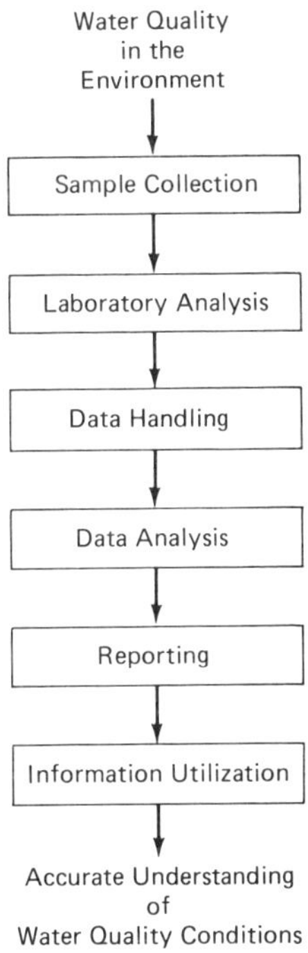

Figure 2.1 The water quality monitoring system following the flow of information.

as the end point.) The last three components in Figure 2.1 can be viewed as the information generation portion of the total system. This information portion has rarely been designed before the initiation of monitoring. When and if the need to analyze the data arises, then this portion of the system is "designed" at that time by the people assigned to "see what the data say."

Since no data analysis methods were established prior to data collection, any efforts to analyze the data are suspected of having some bias. This is especially true when there are disagreements over the findings of a data analysis effort. When such disagreements arise, it is very difficult to address the issues at hand because the arguments do not separate the means of generating information from the results. Arguments over the entire process can render efforts to obtain information at this point almost hopeless.

If, however, the initial, total system design requires that the data analysis methodology questions be addressed before data are collected, then any arguments over data analysis methods are separated from the results of the data analysis exercise. In such an environment it is much more likely that the discussions will focus on the scientific merit of various data analysis methods and not look ahead to the results obtained. This is not to say that data analysis methodologies will always be fixed—invariably data will be useful in ways not foreseen when designing the system; and as better science is developed, it is very appropriate to consider changes in the data analysis methods as amendments to the design. But it does suggest that a monitoring system must be designed in its totality to better define exactly what its purpose is and how it will go about meeting that purpose.

Data collection, which includes the first three components, has historically received the most attention from researchers and monitoring-system designers. Consequently, the ability to quantify the data collection end of a water quality monitoring system is much more advanced than the ability to quantify the information end of the system. In fact, there are many "standard" methods for data collection, laboratory analysis, and data handling that can be utilized. For example, U.S. EPA (1982), U.S. Geological Survey (1977), and American Public Health Association (1985) represent standard approaches to sampling and laboratory analysis.

Fortunately, the information end of the system is receiving increasing attention as efforts are made to determine the most appropriate statistics to use for water quality data analysis. Harcum (1990), Harris et al. (1987), and Helsel (1987) are but three examples of recent work in this area. Reporting has received less emphasis than data analysis, but even here several works are pointing out some of the alternatives (Association of State and Interstate Water Pollution Control Administrators, 1983, and U.S. EPA, 1985a). In both data analysis and reporting, much remains to be done.

Information utilization is the least defined component at this point in time, yet it is probably the most important from an overall information point of view. Since information utilization justifies existence of the entire monitoring program, this text places considerable emphasis on this component of the system and the need to define it as part of the design process.

To sum up, defining the water quality monitoring system as a combination of *data collection* and *information generation* permits a more balanced approach to design and helps the designer achieve a better focus on the information product of the system.

2.2 FURTHER BREAKDOWN OF MONITORING SYSTEM COMPONENTS

Figure 2.1 presented six major components of the water quality monitoring system. Each of these components consists of many specific tasks, each of which must be designed and operated if the total system is to meet its information goals. This section will further define each of the six components to create an outline for designing the operational details of the system. This further definition must, of necessity, be quite brief. Each element of the system will be discussed in more detail later in the book.

2.2.1 Sample Collection

Collecting a sample is the first step in the long flow of information through the monitoring system. Sampling can be as "simple" as dipping a sampling container into a stream or as complicated as drawing a sample up from several hundred meters below the ground surface. The sample can be from the water column, sediments, and/or biomass. Regardless of the sampling situation, there are a number of tasks that must be carried out carefully to obtain a sample that accurately represents the water body. These are summarized in the following list along with some examples of the issues that must be addressed.

Preparation for sampling
- Site access (have legal consents been obtained?)
- Sampling logs
- Equipment availability and maintenance
- Scheduling sample collection (same-time-of-day sampling?)
- Pre-sampling checklist

Sampling procedures (documented? are staff trained?)
- Presampling tasks (e.g., purging a well)
- Sampling protocol (are replicates needed?)
- Sample numbering and labeling
- Sample preservation (physical or chemical)
- Sample transport (fast transport to laboratory required?)
- QC during sampling
- Field measurements (field calibration procedures)
- Field logs

Follow-up to sampling
Filing sampling logs
Equipment cleaning/maintenance
Disposal of chemical preservatives
Has an adequate audit trail been established?

This list of tasks helps in further defining exactly what sample collection involves. Further definition can be obtained from a number of references on the subject. Environment Canada (1983), U.S. EPA (1982), Scalf et al. (1981), Wilson (1980), Dunlop et al. (1977), Mills et al. (1986), Plafkin et al. (1989), and U.S. Geological Survey (1977) provide detail on sampling procedures. Field measurements can vary considerably depending upon the information goal of the monitoring system.

2.2.2 Laboratory Analysis

During laboratory analysis, there are a number of procedures that must be carefully documented and followed if the data are to be consistent and accurate. The following list represents a breakdown of these tasks.

Preparation for sample analysis
Scheduling analyses
Verifying sample numbers
Initiating recording system for sample's results
Initiating sample tracking system through lab
Equipment maintenance
Laboratory analysis
Via standard or well documented method/protocol
Data recording/verification
Coding sheets/data logger
Data verification procedures
Analysis of splits/re-sampling protocol

As noted earlier, laboratory analysis is an area of the water quality information system where considerable documentation is readily available. The American Public Health Association (1985) is probably the most widely known standard for laboratory analysis. As the sophistication of environmental measurements increases, there are additional publications being prepared on the subject of laboratory analysis. For example, the U.S. EPA (1981) describes methods for the analysis of organic chemicals, and Plafkin et al. (1989) describe methods for performing rapid bioassessment.

Quality control within a laboratory is a major element of quality assurance for the entire water quality information system. There are many excellent references on this topic which can be used to define this element of the total system. Taylor and Stanley (1985) is a compilation of papers on the topic of quality assurance for environmental measurements.

2.2.3 Data Handling

The tasks to be performed at this point in the information system are data entry, data storage, data retrieval, and data manipulation for use in data analysis software. These activities can be grouped under the general heading of data handling or data management. For data sets of any appreciable size, a computer and data base software are required to permit ready access to fully documented water quality data for current and future data analysis.

Frequently, for a variety of reasons, the information contained in water quality data is not extracted and utilized. The availability of the data may not be widely known, or its documentation may be ambiguous. (For example, the laboratory method used to determine a nitrate concentration may not have been recorded, or it is not stated whether the units are recorded as nitrogen or as nitrate—there being a four-fold difference between the two). This situation is the inevitable result of poor data handling systems—which often result not just in data ambiguity but eventually in the loss of the data altogether.

There are two fundamental parts of data handling systems. The first is the system used in the laboratory and field to store laboratory analytical results and field measurements. These are the means by which the laboratory and field staff record their own results and keep an audit trail. The second is the general purpose water quality data archive system that makes all the data available, often from various sources, to those who wish to perform data analysis and write reports. These data may include results from more than one laboratory and field party. They may even store results obtained by different agencies.

Because there are many good laboratory data and field data recording systems in use, they cannot all be compatible with the general purpose water quality data archive system. It is therefore a mistake to try to fuse the two components into one. There will be little purpose served by trying to force unwilling laboratory and field staff to change good, familiar methods of data recording. But it is essential that an efficient, friendly interface be built between whatever laboratory and field data recording systems are used and the general purpose water quality data archive system. Then, the data may be accessed using common conventions.

There is a lack of availability of existing general purpose water quality data archive systems. Far too often the data are left on sheets of paper or on an ad hoc computer storage system. In many cases this leads to the above-noted problem that bedevils water quality monitoring: nonexistent, or at best ambiguous, documentation. A generalized data storage system must be designed carefully to ensure that the data are stored unambiguously and are secure. It may be possible to provide for this using general purpose database software such as dBase III or Lotus 1-2-3, but the design of the data management system should incorporate the many essential features described in Appendix B.

The general purpose system used by many government agencies in the United States (STORET), while having many admirable features, is mainframe-based with the access and documentation problems that go with such large systems. Taylor (1970) and Fredrick et al. (1989) describe early and current efforts to develop and upgrade the STORET system. Similarly, the Canadian national system (NAQUADAT) is being upgraded but is still a mainframe system (Brooksbank 1989). In the UK most water authorities have used WATER ARCHIVE (a mainframe-based system), while in New Zealand a system, called AQUAL, based on minicomputers has recently been completed (McBride, 1989).

At present the choice of which data handling system to use will depend upon many factors. In almost all cases, considerable effort will have to be invested in coordinating data handling via a mainframe or in adapting a general purpose data management system for a smaller machine. The size of the monitoring system will dictate whether a large system is needed or whether a smaller, general purpose system can suffice. Personnel support for the operation of the data base system can also influence selection. The larger systems will require more computer expertise to operate while the smaller, general purpose systems can be operated without specialized computer expertise after they have been established.

Operation of the data handling portion of the water quality information system will require regular data entry either manually or by direct read out from the laboratory. There must be means available to check that the data are entered correctly.

2.2.4 Data Analysis

Software packages used to graphically present or statistically analyze the data are the focus of this portion of the water quality information system. Such package(s) must interface with the data storage and retrieval system and provide results in a form suitable for inclusion in reports.

Statistical analysis procedures could be incorporated directly into the water quality data archive system, but it is sometimes simpler to keep them separate but with a clearly defined and efficient interface.

There are a number of commercial data analysis software packages for desk-top machines that can be utilized within a water quality information system. However, they generally do not have features for handling censored data so common in water quality monitoring (i.e., the "less-than" and, less commonly, the "greater-than" data). Nor do they have straightforward procedures for the types of statistical analyses that are being suggested for water quality data (e.g., see Gilbert, 1987, and chapter 6 of this book). Loftis et al. (1989b), and Appendix B describe a computer software package developed specifically for graphical and statistical analysis of water quality data. Gilbert (1987) and Lettenmaier et al. (1982) have both produced trend analysis programs (called TREND). The U.S. Geological Survey has also produced software (Alexander, Schertz, and Ohe, 1989). In Great Britain a recent product, called AARDVARK, has been released (J. Ellis, Water Research Centre, Medmenham, England, personal communication). Other, more general purpose packages, are STATgraphics, SAS, SYSTAT, MINITAB, and DATADESK. Such data analysis packages should ideally interface with word processor/desk-top publishing software to facilitate report writing.

2.2.5 Reporting

Regardless of the automation of the data analysis phase of the information system, interpretation of the results relative to the system's information goals will be needed. This interpretation will require staff time which must be built into the work loads of the monitoring system personnel. This can be difficult since reporting is not a continuous effort (such as sampling and laboratory analyses). However, it is critical that when a report is due, the required staff time is available to prepare it.

The frequency of reporting, the formats of the reports, and the distribution and auditing of the reports are all important aspects of the design of the water quality information system. This aspect for fixed-station monitoring is often overlooked in the initial design and, yet, it is critical to the success of the system meeting its information goals. For special surveys reporting is generally included because it is the last phase of the total survey.

Reporting the results (data and information) obtained from a water quality monitoring system is, today, as much an art as a science. However, a few general guidelines can be stated. First, data lists should be restricted to appendices or to separate data reports. The appendices or data reports can then be referenced. Reports should, in general, become less technical as one moves from the (frequent) technical reports to the (less frequent) public reports. Graphical displays should be used in all reports as much as possible. Careful consideration should be given to developing a water quality "index" for reporting purposes. Such indices can take on many forms—from a single, highly representative variable to a composite index representing a collection of a number of variables.

Reporting has received increasing attention in recent years as efforts have intensified to generate more meaningful water quality information. The evolving 305(b) guidance documents issued by the U.S. EPA are one example of the efforts to improve reporting. The Association of State and Interstate Water Pollution Control Administrators (1984) present a national assessment of water quality, demonstrating a number of means of conveying water quality information, as do Smith et al. (1987) and Wolman (1971).

2.2.6 Information Utilization

When a report from a water quality monitoring system is published, it is distributed to the identified information users. At this point in the total information system, the question of the actual use of the information for management decision making comes to the fore. Is the information from the monitoring system actually being used to help make management decisions? Such use, identified in the initial design, must be documented in some fashion to ensure that the information system is fulfilling its information objectives.

If the information is to account for the agency's impact on water quality, are the general water quality trends being used to do such accounting? If the information is to verify standard compliance within an agency's jurisdiction, are the policy makers and agency staff informed of where the standard compliance problems are located?

To ensure utilization of the information, each report that is issued can have an assessment form in the front which the information user is to complete and return. This should help those running the water quality information system to ensure that their information is appropriate and being used.

2.3 SUMMARY

By defining the water quality monitoring system as an information system with balanced data collection and information generation portions, it is now possible to consider how one would approach the design of such a system. The design must account for each of the six components of the system and be quantitatively focused on the information goals.

When a total water quality monitoring system is being designed with the informational focus advocated in this book, the role of statistics in converting sample results into general knowledge about the behavior of water quality in the environment cannot be overlooked. Use of statistics requires a new way of thinking on the part of many people involved in water quality monitoring. The next chapter is devoted to getting us started in the right direction.

3

THE ESSENTIAL ROLE OF STATISTICS

Monitoring systems, in producing information on water quality, rely fundamentally on data obtained from sampling. But users of the information to be produced from a water quality monitoring program are not interested in samples *per se*. Their interest is in the state of, and changes in, aquatic systems—including the water, the sediments, and the aquatic communities they support. The questions users have, which a monitoring program is funded to provide answers to, are typically of a rather general nature, such as:

1. What are the average water quality conditions in an area?
2. What are the extreme water quality conditions in an area?
3. Are water quality and effluent standards being met?
4. How different are the aquatic communities and water quality in separate regions (e.g., upstream and downstream of an outfall), and why are they different?
5. Is any trend in water quality conditions a cause for concern?

These questions can only be answered by taking *samples* and by making *inferences* about the water quality using the measurements obtained from those samples.

In designing sampling programs to provide such inferences it must be noted that samples are usually only a tiny part of the water body. Also, in the case of biological monitoring, excessive sampling may seriously disturb or destroy aquatic communities, and this must be avoided.

Consider, for instance, a sample volume of 1 L taken once in a month from a stream, and in that month the monthly mean rate of flow was 10 m^3/s. The proportion of that water sampled is about 4×10^{-11}. Or, if on a sampling occasion 10 sediment/biomass samples are taken from a sandflat with an area of 10,000 m^2 using a sediment sampler with a diameter of 10 cm, the proportion of the sandflat sampled is about 8×10^{-6}. If these tiny samples are to represent the vast volume of the aquatic system about which questions are to be answered, considerable care will be needed in their collection, analysis, and interpretation. At every stage there will be some level of uncertainty: about the representativeness of the samples; the potential for error in taking and analyzing the samples; and in the interpretation of the results.

Such uncertainty cannot be avoided altogether, and many aspects of science are needed to see that it is kept to reasonable levels. Some examples are: knowledge of and training in sampling techniques that avoid sample contamination; use of good laboratory procedures and tests; and understanding mixing phenomena and the functioning of aquatic communities. And running through all these is the subject of statistics, which has been developed specifically for analyzing and accounting for such uncertainty.

The proper use of statistical methods, *both in designing a sampling program and in interpreting the results obtained,* offers the huge advantage of consistent, repeatable procedures as a means of accounting for and quantifying the uncertainty. Different investigators, using the same statistical methods on the same data should get the same result. So, the all-too-common conflict of subjective opinions about just what data ought to be collected and how they should be analyzed can be minimized. Such procedures can be used in conjunction with data on costs of sampling so that money and effort can be spent wisely (Mar et al. 1986).

Of course some subjectivity will always remain—in deciding what variables to measure, what methods to use, what level of uncertainty can be tolerated, and how the answers to be obtained by the statistical methods relate to the more general questions of interest to the user (i.e., questions 1—5). But it would be extremely unwise to proceed with a sampling program ignoring the implications of the statistical methods on its design and interpretation.

3.1 STATISTICAL UNDERSTANDING NEEDED

Before statistical methods are applied in the design and analysis of water quality monitoring programs, an understanding of two important features of water quality data are required. The first is an understanding of the *types and scales* of water quality data. These influence what methods can be used. The second feature is an understanding of the relationship of *populations, distributions, and sampling regimes.*

Then, to answer the questions to be addressed by the monitoring program, a monitoring system designer uses methods of statistical inference. Where the question of interest concerns *quantifying* average or extreme water quality conditions or changes therein, methods of *estimation* apply. When the question of interest concerns comparisons of water quality with a standard, over space or over time, *hypothesis testing* methods also apply.

The design of water quality monitoring systems must include clarification of the four topics highlighted in the preceding paragraphs—types and scales of measurement; populations, distributions, and sampling regimes; estimation; and testing.

Whenever the task of monitoring system design is undertaken, it is assumed that the designer has a familiarity with basic statistical theory and practice. Sanders et al. (1983) and, particularly, Gilbert (1987) both contain information on statistics as they relate to water quality. There are a number of basic statistical texts that are very useful: Bowker and Lieberman, 1972; Sachs, 1984; Hines and Montgomery, 1980; Miller and Freund, 1985; and Zar, 1984.

The word "sample" is used ambiguously in this text as it is in most efforts to measure water quality. "Samples" may refer to water quality data obtained by any of three methods: *in-situ* probe measurements; site observations; or laboratory analyses on volumes of water, sediments, and biomass. A "sample" may also refer to either a single datum so obtained or to an entire collection of data (which is the usage of the word in statistical literature). The context should make it clear which meaning is intended.

The terms "variable" and "parameter" are used in the statistical context. A water quality "variable" is the characteristic being measured (such as pH, temperature, dissolved oxygen, and chlorophyll *a*). As Steel and Torrie (1960) note, a variable is a characteristic that shows variability or variation. A specific measurement is referred to as an "observation" recorded, for example, on a given date. In the jargon often accompanying monitoring system design, such variables are referred to as "parameters." This causes problems when, in using statistics in the analyses of data, the quantities such as the "mean" or "standard deviation" are referred to as parameters. The word "parameter" is used herein for characterizing populations, and "statistic" is used in the case of samples.

This chapter provides the reader with the general understanding necessary to be able to use the data requirements of the appropriate statistical methods to state explicit design objectives. Such objectives are a key aspect of design. So, from the general, and therefore ambiguous, questions to be answered by the monitoring program, explicit numerical statements of objectives can be formulated that lead to a relatively straightforward design procedure. This point has been made by several authors: Price, 1987; Whitfield, 1988; Mar et al. 1986; and Ward et al. 1986.

3.2 TYPES AND SCALES OF WATER QUALITY DATA

Water quality conditions are measured by variables of two types: *continuous* or *discrete*. They are also measured or reported on different *scales*. These are fundamental distinctions, dictating which statistical methods may or may not be used. Table 3.1 gives a summary of the various combinations possible with water quality variables (including biological variables). As one can see from the following discussion, when the authors refer to a measurement or reporting scale, they do not mean the monitoring scale (the time or spatial scales over which data are collected).

It should be noted that the view of water quality by various people is strongly influenced by the temporal and spatial scales of monitoring. For example, the degree of change in water quality which represents a "significant" trend, as opposed to random fluctuation, depends on whether the time horizon is one or a hundred years.

Table 3.1. Examples of types and scales of water quality variables.

SCALE OF MEASUREMENT (DATA MAY BE RECORDED ON A LOWER SCALE)	DATA TYPE			
	DISCRETE		CONTINUOUS	
	PHYS./CHEM.	BIOLOGICAL	PHYS./CHEM.	BIOLOGICAL
Interval	. . .	Organism numbers	Most variables (pH, DO, . . .)	Organism size
Ordinal	Visual clarity	Organism size class	. . .	. . .
Nominal	Water hue	Organism name	. . .	. . .

3.2.1 Data Types: Continuous Versus Discrete

Most physical/chemical water quality variables and some biological variables are continuous (see Table 3.1). Such variables may, in theory, take an infinite number of numerical values, even if they are confined to a range (e.g., pH). Of course, when a continuous variable is measured, the limited precision of measuring instruments and analytical tests mean that there are only a finite, though large, number of values the result can take, so the data will actually be discrete. Nevertheless, the term ''continuous'' is used to describe such variables. This is because were the variable to be measured continuously through time, the results may be shown as a smooth line. The many tiny steps actually inherent in the data are not discernable to the eye.

Discrete variables (most biological data) can take only a finite number of ''values''—in theory and in practice. These ''values'' may be numerical or an alphabetical code. Were it possible to record such variables continuously, the results would fall on a stepped line, not a smooth line. Note that at the low end of their range some continuous variables are reported as discrete. When a laboratory measurement has such a small value that the laboratory is not confident in the value, the results are often reported as ''less than detection limit.'' Such data are referred to as ''non-detects'' or as being censored (Porter, Ward, and Bell 1988).

Occasionally data are also recorded as ''greater than'' a limit. Examples would include results of a BOD_5 test using insufficient dilution when there is no more sample left to repeat the test at a higher dilution or a plate count of microorganisms which gives a ''too-numerous-to-count'' result.

3.2.2 Scales of Measurement and Reporting

Water quality variables, continuous or discrete, can be measured on a number of scales. The first, and ''highest,'' is the *interval* scale. The fundamental feature of this scale is that size of the intervals between data on the scale has a clear meaning. Measurements on this scale are always numbers. All continuous data and some discrete data are measured (though not always reported) on this scale. Actually, in the definition of scales in common usage there are three similar interval scales (Zar 1984), one of which is also called the interval scale. These are:

- the ratio scale (most physical/chemical data except non-detects and ''greater thans,'' biological counts)—data have an absolute zero so

that ratios have a consistent interpretation, e.g., 20 ppm is twice 10 ppm, and 50 ppm is twice 25 ppm;
- the interval scale (temperature in 0°F or 0°C, transformed variables like pH)—these do not have an absolute zero, so that ratios do not have clear meaning, e.g., 20°F is not twice 10°F; and
- the circular scale (time, compass direction)—periodic interval data.

Some discrete data are measured on the lower *ordinal* scale or, more usually, on the lowest scale—the *nominal* scale. Data on these scales are always discrete (see Table 3.1) and may be represented as numbers or as alphabetical characters.

Ordinal data imply an ordering (i.e., ranking)—they may be arranged from smallest to largest or from worst to best. The only interpretation of the interval between data on this scale is relative, e.g., *A* is greater than *B*.

Few water quality characteristics are actually measured on this scale. However visual observation of water clarity is one example, where an order is implied in observations such as "clear," "slightly muddy," "muddy." Nondetect data are reported (though not actually measured) on this scale. Another case is where limitations of methods for measurement of a continuous variable mean that only a few discrete numerical values can be attained (e.g., low concentrations of organic-nitrogen).

The usual reason, however, for the occurrence of ordinal data is where data that could have been expressed on the higher (interval) scale are deliberately manipulated (by ranking) so that they are expressed on the ordinal scale. There can be advantages in doing this, even though some information may be lost in the ranking process. An obvious advantage is that where the data for a continuous variable contain non-detects, ranking the data reduces them all to the same scale. The rank transformation is used frequently in nonparametric statistical procedures, a class of methods becoming increasingly popular in water quality because of its broader applicability than classical methods based on the normal distribution.

Nominal data merely record categories or attributes of data, and do not imply a ranking. They may record several categories (e.g., of water hue—blue, blue-green, green-yellow, . . .), but often only two are recorded (especially presence/absence for biological data, exceedance/non-exceedance for physical/chemical data), in which case the data are said to be *dichotomous*. Once again, data on the higher scales can be reduced to the nominal scale. For example, continuous river pH data (interval scale) can be reduced to a dichotomous nominal scale by recording merely whether each value exceeds or does not exceed some limit or range. Information is lost in this process too, of course.

The gulf between data on the nominal and ordinal scales is much wider than that between the ordinal and interval scales. This is because the ordinal and interval scales at least have the relative size of data as a common feature—the nominal scale does not.

3.2.3 Implications of the Data Type

The type of data dictate what kind of distributions are used in the appropriate statistical methods. Continuous data, which are of course confined to the interval scales (see Table 3.1), use statistical methods that are based on continuous distributions. Typically this is the normal (sometimes called Gaussian) distribution. For skewed data, which are common in water quality, the lognormal distribution is often used. Gilbert (1987) gives an excellent account of this distribution. Other continuous distributions (e.g., Weibull, gamma or beta—see Gilbert 1987:155) are used occasionally.

For routine applications in water quality and other types of environmental monitoring, when data are often not normally distributed, data are transformed to the ordinal or nominal scales, and nonparametric methods are used. There have been many successful implementations of nonparametric methods in water quality statistics in the last decade (e.g., Helsel 1987).

Discrete data, which can be on any scale, typically use statistical methods based on discrete distributions. These are the Binomial, Poisson, Negative Binomial and multinomial distributions (Freund 1971, Elliott 1977).

There are exceptions, especially when interval scale data containing non-detects have those data replaced by some value (e.g., one-half the detection limit). The data are then a mixture of continuous and discrete types, but parametric methods can sometimes be used (Zar, 1984:138).

3.2.4 Implications of the Data Scale

It often comes as a surprise to learn that, with one exception (estimation with nominal scale data—see section 3.4), there are statistical methods that apply to data on the lower scale. For example, "Cochran's Q test," based on the chi^2 distribution, has been applied successfully to presence/absence biological data (Pridmore 1985). It is important that the applicability of statistical methods over all data scales be well understood, particularly because many of the data collected in biological monitoring programs and in field observations are on the lower scales.

There are two points to bear in mind. First, a method that works on data from a lower scale will work on all higher scales, but the converse is not necessarily true. For example, nominal scale methods—which are based exclusively on chi^2, binomial, or multinomial distributions—can also apply to data on the ordinal or interval scales. But any method where relative ordering is a requisite cannot work on the nominal scale.

Second, the questions that can be asked of data get more specific the higher the data scale. That may be seen as an advantage, but on the other hand methods to address such specific questions generally require more specific assumptions. This is often a disadvantage, particularly if the assumptions cannot be met!

3.3 POPULATIONS, DISTRIBUTIONS, AND SAMPLING REGIMES

Actual water quality conditions changing with time at a site or changing through space at a particular time, can be shown as a trace on a graph. For example, suppose that the cross-section-average pH (interval scale) at a river site is measured using a continuous monitoring system, as

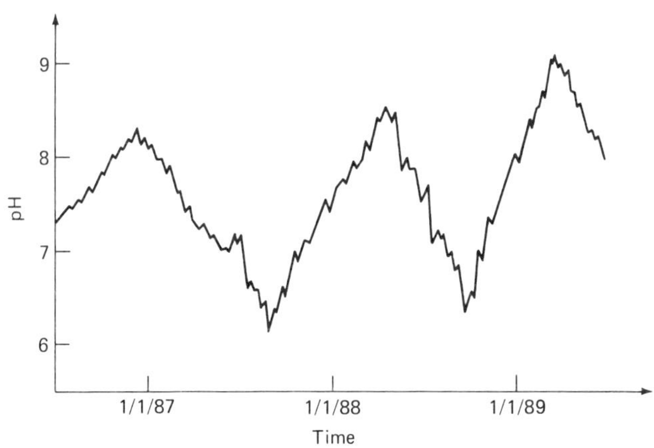

Figure 3.1 Hypothetical time trace of pH.

depicted in Figure 3.1. Because pH is a continuous variable, the trace is a smooth line. The graph shows the entire *population* of the water quality variable pH at this site for the three years it was recorded.

3.3.1 Summarizing the Population—Distributions

Such a graph shows an enormous quantity of data and gets ever longer as sampling proceeds. It will often be necessary to summarize the data

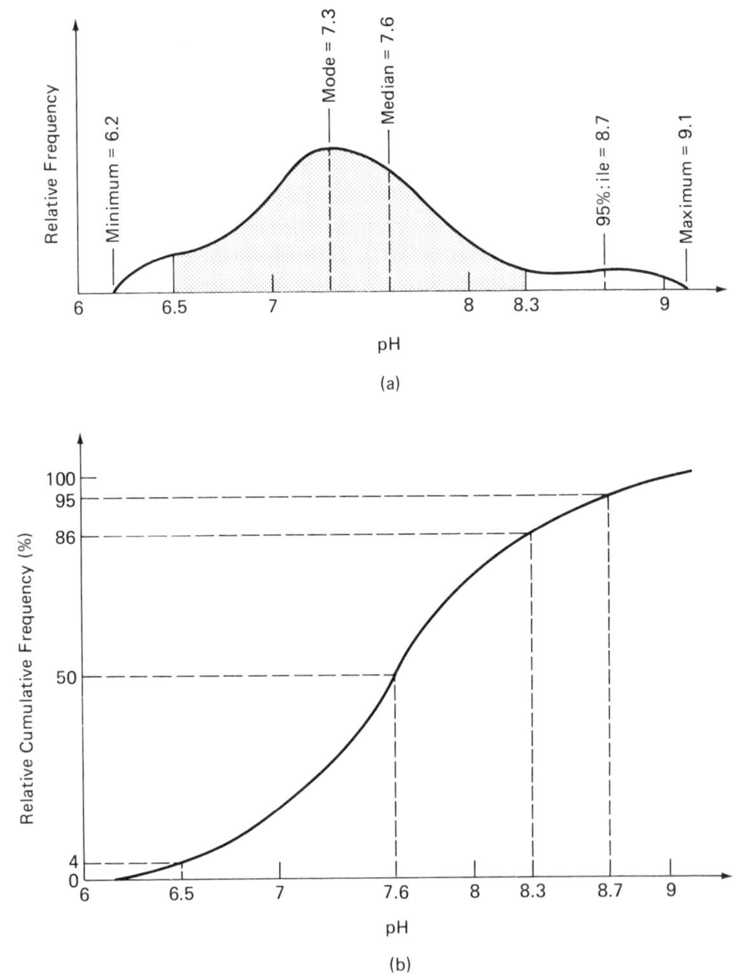

Figure 3.2 Relative frequency distribution (a) and cumulative frequency distribution (b) for the pH data in Figure 3.1.

using a smoothed *relative frequency distribution* shown in Figure 3.2*a* or its discrete counterpart, a histogram. When this is done, a number of features of Figure 3.1 are retained and even highlighted (Miller and Ellis 1986).

The graph has the advantage of showing the shape of the distribution of pH in the sampling period, with the minimum, maximum, and mode being particularly evident. This graph also shows the fraction of time (i.e., the frequency) that the pH was between two values (say 6.5 and 8.3). This fraction is the ratio of the shaded area to the total area under curve. This of course is tedious to compute from Figure 3.2*a*, but it may be read directly from the smoothed *cumulative* relative frequency distribution (sometimes called an ogive), which may be obtained simply by integrating the relative frequency distribution, as shown in Figure 3.2*b*. This curve shows that the required percentage is 86% − 4% = 82%.

Note that the penalty to be paid in forming this useful summary is that all features relating to the time-order of the data shown in Figure 3.1 are lost in Figure 3.2. For example, Figure 3.2 gives no indication of the seasonality evident in Figure 3.1.

Taken together, the time trace and the frequency distributions give a rather complete picture of the population. The problem of how to make inferences about the population from a tiny sample is avoided, because the entire population has been measured directly. There is no uncertainty, assuming the pH does not vary in the river cross section and was measured accurately (which is not a trivial task in river waters with low buffering capacity—Schroder et al. 1989, Ficken 1989, Smith et al. 1989). Therefore many questions that may be asked of these data can be answered directly.

3.3.2 Implications for the Sampling Regime

The luxury of a continuous trace of a water quality variable is rarely encountered. In water quality monitoring a typical sample size is a mere 50, and is often smaller. That being so, samples can be regarded as providing an approximation or an estimate of the relevant time trace and of the frequency distribution. The population frequency distribution is now unknown, and referred to as the *probability density function*. That is, for a finite sample size the idea of measured frequency is replaced by estimated probability. The need for accuracy and precision in estimating the characteristics of a probability distribution will place requirements

For example, consider the situation where pH is measured at an average frequency of once per month at the river site for a period of three years (sample size = 36), and it is desired to estimate the time trace or frequency curve for that whole period. If it is desired that the estimates be as accurate and precise as possible (i.e., to be unbiased and have minimum expected variance), the sampling should be *random* throughout the whole period. That is, any point on the trace should have an equal chance of being selected. If sampling is not random, there is a risk of the results being biased.

If, for instance, the sampling site is downstream of a dairy factory that often discharges acid-wash water from boiler cleaning operations late on Friday, at about 4 P.M., the discharge can cause a depression in pH in late afternoon, which of course would be measured by a continuous monitor. But if samples are collected only at about 1 P.M. on weekdays (perhaps because of practical problems of work scheduling, transportation, and safety), the depression is never measured. Therefore certain portions of the trace and of the histogram will be biased upward.

However for some purposes the idea of estimating the trace and distribution for the whole period can be abandoned, and interest can be confined to the period around 1 P.M. on weekdays. Then the estimates *for that period only* are not biased.

In practice, sampling is never random—there is always some systematic component. This is the inevitable consequence of the practical realities of running a sampling program. Managing equipment and staff availability, costs (especially overtime rates for extra hours), and laboratory throughput can be particularly difficult. But it must always be kept in mind that routine sampling may miss some regular variation in the water quality variable and therefore be biased.

Gilbert (1987: chapter 3) gives an excellent discussion of sampling regimes and their consequences. Several authors have noted that in "impact studies," simultaneous sampling at control and impact sites is often necessary. Green (1979: chapter 3) discusses optimal design for assessing the impact of an effluent discharge on a river, where samples are taken upstream and downstream of the discharge both before and after impact occurs. The upstream site is the "control." In an influential paper, Hurlbert (1984) states objections to this design, saying that all such a design can do is to demonstrate differences and not impacts. Stewart-Oaten et al. (1986) have shown that most of the objections are overcome if the control and impact sites are sampled simultaneously, using two-sample tests on the differences between simultaneous samples. Other authors have also advocated simultaneous sampling at control sites (Ponce, 1982; Millard and Lettenmaier, 1986; Loftis, Harris, and Montgomery, 1987). Control and impact sites may not always be easy to find, especially in estuaries or in highly impacted aquifers affected by many sources of contamination.

3.4 ESTIMATION

The general questions 1 and 2, noted at the beginning of this chapter, are the starting point when designing a water quality information system to estimate average and extreme conditions. It does not necessarily follow that the only parameters of interest are the mean (for the average condition) and the minimum or maximum (for the extreme condition). Other summary statistics such as medians or percentiles may be more appropriate.

Note that merely estimating the value of the chosen summary statistic (i.e., a point estimate) is rather unsatisfactory. It is highly desirable to have some measure of the uncertainty or precision of that estimate. This is achieved by computing a confidence interval (i.e., an interval estimate) around the estimated statistic. One should be confident that such an interval contains the true statistic. Note too that estimates can only be made for these parameters using data on the ordinal and interval scales. Nominal scale data do not have a sense of relative size so notions of ''average'' or ''extreme'' have no meaning.

3.4.1 Choice of Statistics to be Estimated

Referring to Table 3.1, three cases must be considered:

(1) Continuous interval scale data. To translate questions 1 and 2 into an explicit choice of summary statistics to be estimated, the general shape of the distribution from which one is sampling must be determined. Of course this distribution will be continuous or discrete, depending on the type of data being acquired.

In this age of widely available computer technology, the first step should be to view the data. That is, use Exploratory Data Analysis (EDA), provided by modern statistical packages. Velleman and Hoaglin (1981) give an excellent account of EDA methods. Recently, specialized packages (e.g. WQSTAT developed by Loftis et al. 1989b; and AARDVARK developed by the Water Research Centre, Medmenham, England) have enhanced EDA for water quality data. Such viewing should make use of the time trace, frequency distributions, and box plots (which are particularly good for comparing two or more data distributions).

After viewing the data, it may appear that the distribution is quite skewed, and the mean would be a poor estimate of average conditions. Perhaps the median would be more appropriate. The median is generally recommended in water quality data analysis instead of the geometric mean, to avoid the problems with using data transformations (Gilbert, 1987:149). For extreme conditions, it may be that the minimum and maximum are outliers and may be in error, or that the minimum and maximum are only fleeting chance events of no great concern, and that some lesser percentile may be more appropriate. (For example, in England and Wales stream and effluent standards are based on 95th percentiles of water quality distributions over time.)

(2) Discrete interval scale data. For physical/chemical variables, the same consideration as in (1) also arises. For biological variables it should be noted that organisms tend to go through occasional periods of rapid change (caused by recruitment, spates, . . .), so that their temporal distribution is skewed. So to characterize the average population size, in time, the median may be the best statistic. Their spatial distributions may be skewed also, depending on the scale of measurement as determined by the size and placement of samplers.

Aquatic biological populations tend to be very clumped at intermediate scales (Elliott, 1977). In this case, even though data are typically very skewed, it may be appropriate to estimate the mean and its confidence interval. This is because to calculate the total population or biomass in an area, one has to use the mean population in the samples. Any other statistic would produce error.

(3) Discrete ordinal scale data. There is no point in viewing these data. They always have the same shape—just an ordering of data from top to bottom (or bottom to top). So percentiles of the data are used as summary statistics, typically the median and some extreme percentile are most meaningful.

3.4.2 Implications of the Estimate

Methods for calculating the appropriate confidence intervals have strong implications for the decision on the required sample size to be taken as part of the design of a sampling program. Consider the simple case where random samples are taken from a normal distribution with unknown variance, and a confidence interval is found for the estimate of the mean. This estimate is the sample mean, for which the confidence interval is

$$\bar{x} \pm w : w = t_{\alpha/2,\nu}\, s/\sqrt{n} \tag{3.1}$$

where: $\bar{x}$ is the sample mean; w is the confidence interval half-width; 1-α is the *confidence coefficient*—the before-sampling probability that the confidence interval will contain the true mean when constructed from a random sample of n observations (after sampling the interval either does or does not contain the true mean with probability one, but we cannot know which); ν is the degrees of freedom (in this case, n-1); s is the sample standard deviation; and t is the α percentage point of the Student's t distribution (the value of t to the right of which the tail area is α).

This formula can be inverted to give a simple equation for the sample size necessary to estimate the mean with a desired confidence. It is:

$$n = s^2[t_{\alpha/2,\nu}]^2/w^2 \tag{3.2}$$

So once a confidence coefficient and a desired ratio of the standard deviation to the half-width are specified, the sample size can be calculated. Therefore, in the design process, once appropriate values of these two quantities have been agreed upon, the required sample size can be calculated. In this way, explicit design criteria can be formulated.

Further detail on sample size calculations for various cases appears in Appendix A.

3.5 HYPOTHESIS TESTING

The general questions 3, 4 and 5 (see page 28) are the starting point when the sampling program is designed to provide data for making comparisons. Comparisons rely upon the field of hypothesis testing which, for ordinal and interval scale data, is very closely allied to estimation. The difference is that in making a comparison, an extra piece of information is used—the estimate has to be compared with some reference, for example a water quality standard, as in question 3. But whereas estimation works only on ordinal and interval scale data, hypothesis testing works also on nominal scale data—so tests can apply to lists of biological species indicating presence or absence, for example.

While hypothesis testing is usually presented in an apparently straightforward manner, the subject is still fraught with conceptual difficulties and some controversy. This is particularly true in the behavioral sciences and is summarized by Morrison and Henkel (1970) and, more recently, by Oakes (1986). The following text material attempts to clarify the issues, but the final word has not been written on this topic as it relates to water quality monitoring.

3.5.1 Conceptual Framework

In hypothesis testing a *null hypothesis* (denoted by H_0) and an *alternative hypothesis* (denoted by H_A or sometimes by H_1) are stated. Then a test is devised and used on data collected to decide whether or not H_0 should be rejected. If H_0 is rejected, then H_A is accepted and vice versa.

There are no hard and fast rules for how hypotheses are to be selected, but in water quality the null hypothesis would usually be one of no change or no difference (e.g., in question 4 H_0 would be phrased to say that there is no difference in a water quality variable between regions, or more specifically that the means or medians of the distributions in the two regions do not differ).

The business of hypothesis testing is then to find a way of choosing between the two hypotheses. This is achieved by performing an *hypothesis test* on the sample. This involves using a *decision rule* to decide whether to reject H_0. If the rule leads to a false H_0 rejection, a *Type I error* will have been committed. So this rule is devised subject to a maximum allowable Type I error risk, called the significance level and denoted by α, and the test may be known as a "significance test." Because H_0 usually asserts the *status quo,* α has been called the "producer's risk" and can be characterized as a "false alarm."

There are two types of error in hypothesis testing. In estimation problems there is only one kind of error that could be made—that the true parameter may not fall within the confidence interval. But in hypothesis testing it is fundamentally important to note that because comparisons are being made, another error may occur—even though it is not considered when performing the test. As well as the Type I error, there is the possibility of committing the *Type II error.* A test can lead to the conclusion that the null hypothesis is true when in fact it is false. The risk of this happening is denoted by β. This error has been called the "consumer's risk" and can be characterized as "slipping through the net."

The way this fits together is shown in Table 3.2, which also shows the definitions of the terms "confidence level" ($1-\alpha$) and "power" ($1-\beta$).

Table 3.2. Four possible outcomes of hypothesis testing and their probabilities.

	"TRUTH"	
TEST RESULT	H_0 TRUE	H_A TRUE
H_0 true	$1 - \alpha$ (Confidence level)	β (Type II error risk)
H_A true	α (Significance level) (Type I error risk)	$1 - \beta$ (Power)

In environmental monitoring the risk of committing either error should generally be kept low so that whichever conclusion is reached has a high probability of being correct. Considerable money and effort are now spent on waste collection, treatment and disposal, and not too many false alarms can be tolerated (i.e. α should be minimized). But it is also desired to minimize the Type II error risk, β (i.e., maximize power). The aquatic environment does need protection from abuse, and there is a desire to not let pollution slip through the net.

Because the sample size cannot be set independently of the error risks, design will usually involve a tradeoff between the two risks. Keeping both α and β small would require that only large differences could be detected or a large sample size is needed.

Surprisingly, little attention to Type II errors is given in standard texts, and they have therefore often been ignored in designing water quality monitoring programs (Toft and Shea, 1983; and Millard, 1987). The reasons date back to the original development of tests in agricultural research in the 1930s (Fisher 1954, 1960), and are not of concern here.

To see how the tradeoffs between Type I and Type II risks may be made, the simplest case will be examined first.

3.5.2 Tests on Nominal Scale Data

Suppose a river water quality standard states a percentile pH standard. For example, the pH should be in the range of 6.5–8.3 for at least 95% of a month. It is desired to test whether the pH was within that range for at least 95% of the month. If it was not, the river standard has been breached. A 10% significance level is adopted, and the sampling frequency is 16 samples per month. If p is the true proportion of the time that the limit was not exceeded, then using a null hypothesis that the discharge does not breach the standard results in a test for:

$$H_0: p \geq = 0.95 \text{ versus } H_A: p < 0.95$$

These hypotheses are both *composite* (they are inequalities), and the test is therefore *one-sided* (or *directional*). The decision rule relies on the remarkable result that, provided sampling is random, the number of exceedances of a limit always follows the binomial distribution regardless of the distribution the data follow (Ellis and Lacey, 1980). The test proceeds by converting the interval scale pH data to dichotomous (exceed/not exceed) nominal scale data.

Using e to denote the maximum allowable number of exceedances of the pH range in n samples, it is possible to allow $e = 2$ exceedances of the pH range in the $n = 16$ samples and still keep the Type I error risk below 10%, even in the borderline case where p is at the critical value of 0.95 (the method by which this result was calculated is given in Appendix A).

Such a result usually comes as a surprise, as this is 12.5% of the sample, more than 10% of the samples. In fact, if a 5% significance level had been chosen this many exceedances are still allowed! (Such results, often counter-intuitive at first, are common with binomial probabilities.)

One may then wonder what is the power of this test to detect a breach. While many words and equations may be invoked to answer this, by far the best way is to view the *Operating Characteristic curve* (OC curve) for this test. Once OC curves are understood, the performance of significance tests can be evaluated at a glance.

The OC curve, as shown for this test on Figure 3.3, is a plot of the probability of accepting the null hypothesis versus all possible values of the true compliance proportion (p). To the right of $p = 0.95$ the river is in compliance with its standard (H_0 is true), because the true compliance proportion is greater than the critical value of 0.95. Therefore, in this region a Type I error could occur. It would occur if by chance three or more exceedances occurred in 16 samples (the decision rule would then lead to a conclusion that H_0 should be rejected). Therefore in the compliance region, the distance *above* the OC curve is the Type I error risk (α). Equivalently, the height of the curve is the confidence level ($1\text{-}\alpha$).

On the other hand, to the left of $p = 0.95$ the river is in violation of

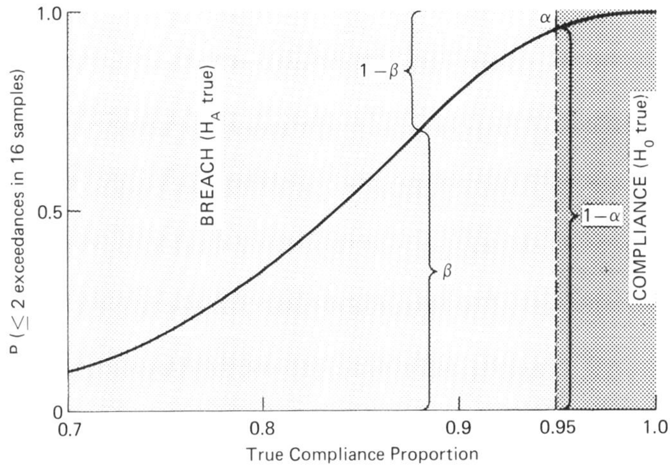

Figure 3.3 OC curve for $n = 16$, $e = 2$.

its standard (H_A is true), because the true compliance proportion is less than 95%. Therefore, in this region a Type II error could occur. It would occur if fewer than three exceedances were drawn in 16 samples (because the decision rule would lead us to conclude that H_0 should not be rejected). The probability of this happening is the height of the OC curve (β). The distance above the OC curve is the power, 1 - β.

Thus the OC curve shows completely all the major features of the test. In particular it shows that the error risks are not constant: they both get bigger the closer p gets to the critical $p = 0.95$ line, at which point $1 - \beta = \alpha$. That is, the power equals the significance level. But in general the Type I error risk is much smaller than the Type II error risk. Even if compliance is borderline ($p = 0.95$) there is only a 4.3% chance of falsely inferring a breach. The true compliance proportion would have to drop to about 0.65 before a similar Type II error risk is faced.

In between $p = 0.65$ and $p = 0.95$, there is a grey area. In other words, the hypotheses are treated *asymmetrically:* the risk of falsely inferring compliance is much greater than the significance level. The reverse would be true if the chosen null hypothesis was that the standard had been breached.

Finally, comparing the OC curves for two different decision rules, sufficient information is available to do the tradeoff between Type I and Type II error risks. For example Figure 3.4 shows the $n = 50$,

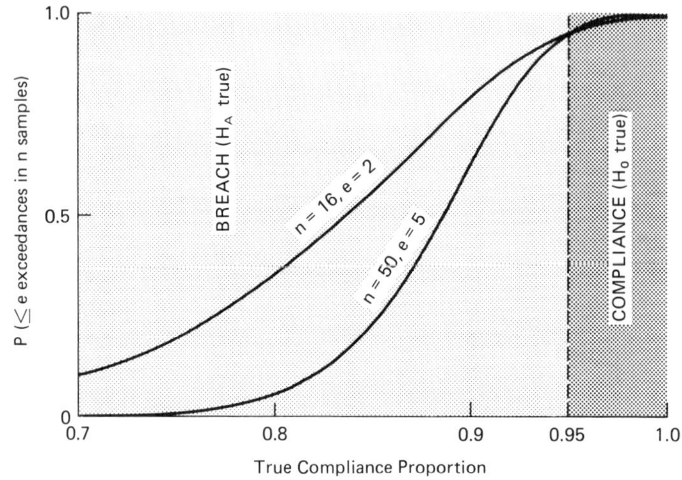

Figure 3.4 OC curves for two decision rules

$e = 5$ OC curve for the original hypotheses. This decision rule also has a maximum Type I error risk less than 5% (because the binomial distribution is discrete, α cannot be made exactly equal to 5%), but the curve does not intrude so far into the breach region. The grey area has been reduced; increasing the sample size has made the test more powerful.

Unfortunately many introductory texts do not show OC or power curves. It seems to be held that they are too difficult to understand (e.g., Howell 1982:152). Yet, they are a valuable aid to interpreting tests. Ferris, et al. (1946) called them ". . . a necessary adjunct to the proper interpretation of the common tests of significance."

In introductory texts (for example Neter et al. 1979) OC curves are sometimes introduced generally, without giving specific cases. Sokal and Rohlf (1969) introduce power curves rather than OC curves. Wallis and Roberts (1956) discuss OC curves for nominal scale data only. Other texts do give at least some OC curves. Bowker and Lieberman (1972) and Miller and Freund (1985) present them for nominal scale data in the related topic of acceptance sampling. These texts also present OC curves for parametric tests on means and variances, as do also Hines and Montgomery (1980) and Montgomery (1985). Larsen and Marx (1986) present power curves for *t*-tests. Zar (1984) also presents many procedures for calculating power.

In the behavioral sciences, Cohen (1988, first published in 1969) presents many power calculation procedures, some of which are repeated in Howell (1982) and Welkowitz, Ewen and Cohen (1982). OC curves and other power computation procedures do appear in more specialized advanced texts such as Beyer (1968) and Pearson and Hartley (1966).

When a significance test results in rejection of the null hypothesis, it is common to call the result "significant." This sounds important, and rigorous. All too often, it is not. For instance, the more powerful decision rule in the earlier example ($n = 50$, $e = 5$) will detect more breaches than will the $n = 16$, $e = 2$ rule. So with the former, more "significant" results will be obtained. It follows that "significant" results should only be compared for tests having the same OC curves (identical sample size and significance level). This holds true for all hypothesis tests including analysis of variance, regression, and goodness-of-fit.

3.5.3 Tests on Ordinal or Interval Scale Data

Tests on ordinal and interval scales can be seen as an extension of estimation problems, which work only on ordinal and interval scale data. Consider the case where the population mean pH is to be compared for two rivers. Because it is not usually known in advance which population mean is the greater, a simple null hypothesis (an equality) is appropriate:

$$H_0\text{: } \mu_1 = \mu_2 \text{ versus } H_A\text{: } \mu_1 \neq \mu_2$$

where μ_1 and μ_2 are the population means. Because H_0 is simple, the test is "two-sided." For normally distributed data, the appropriate test is the t-test, described in most statistics texts. The test works by rejecting H_0 if the appropriate confidence interval for μ_1-μ_2 does not contain zero. Some operating characteristic curves, for $\alpha = 0.01$ and 0.05, are shown in Figure 3.5.

Most of the features of these curves are the same as for the binomial test. The power can be increased by increasing any or all of: (1) the sample size; (2) the significance level; and (3) d, which is a relative measure of the true difference of means. But now, the "H_0 true" region has shrunk to a line, because the null hypothesis is simple. A moment's reflection shows that, for a two-sided test on continuous data, *the null hypothesis can never be true*. The probability of a continuous variable taking a particular value is zero (Freund, 1971:104). Even for discrete data the probability is very small. As a consequence the null hypothesis will always be rejected more frequently as the sample size increases.

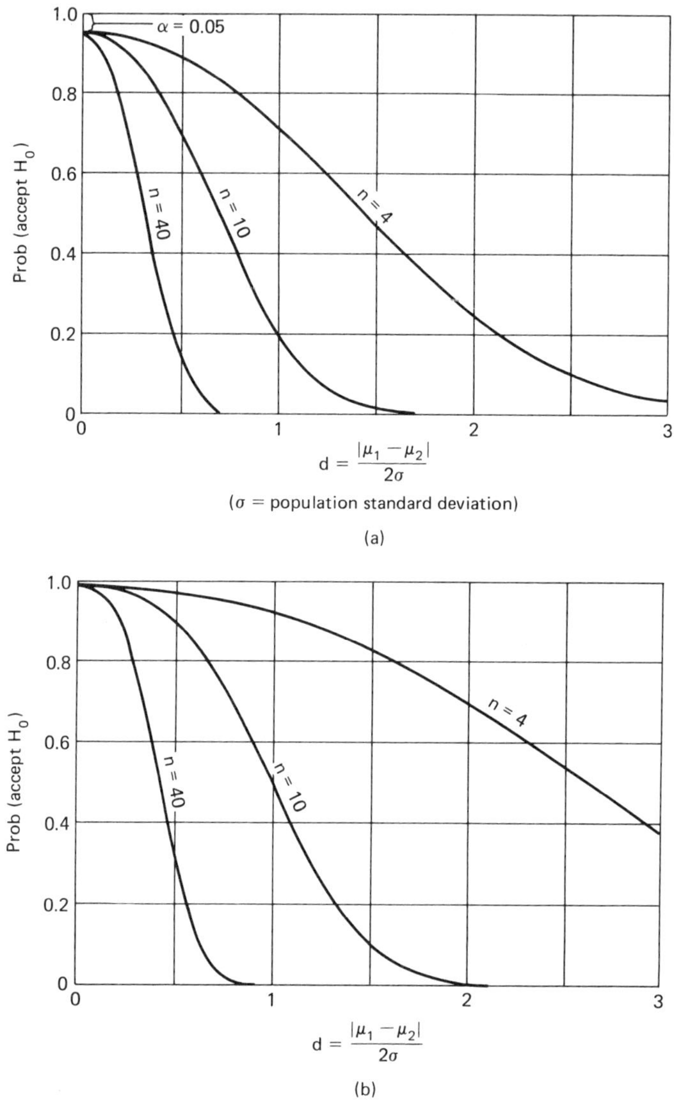

Figure 3.5 OC curves for two-sided t test for $H_0{:}\mu_1 = \mu_2$. (a) $\alpha = 0.05$; (b) $\alpha = 0.01$

Now the question of interest is not really whether the null hypothesis is true in an absolute sense but rather (as questions 3–5 imply), how likely is it that the true difference is greater than some specified value? Some true difference can always be expected. What is desired is that the test should have a good probability of detecting a difference if it is of a certain magnitude.

The reason that an "unrealistic" null hypothesis is used (Cochran and Cox 1957:5) is that it makes the test simple, using the *t* distribution to describe the distribution of the test statistic under the null hypothesis and to define critical values for acceptance or rejection. Otherwise, the more complicated non-central distribution would have to be used (Larsen and Marx, 1986).

Care is needed in interpreting tests of simple hypotheses. Again, the OC curve comes to the rescue. Consider the relative true difference, *d*. In designing a sampling program the value of *d* can be specified—a value which should be detected at a given power. Then the required sample size can be read directly from the OC curve.

Having set *d* and power, the result of a test can be interpreted in the following way. Say the test, with a 5% significance level, is to detect a value $d = ½$ with power $1\text{-}\beta = 90\%$. Referring to Figure 3.5, the required size of each sample is about 40. If *d* is in fact close to zero with this sample size, there would be only a 5% chance of the test rejecting H_0 and so concluding falsely that *d* is greater than ½. If *d* was in fact at least equal to ½, there would be at most a 10% chance of the test failing to detect that. But if *d* was in fact somewhere between ½ and zero, there is still a grey area where the null hypothesis may or may not be rejected. This simply cannot be avoided.

Statistical texts often introduce the concept of a null hypothesis by saying something like "suppose a lightbulb manufacturer claims that the average lifetime of the bulbs is exactly 400 days, and this claim is to be tested." In fact no manufacturer should make such a claim since it cannot be right, the probability is zero that the average lifetime is exactly anything when measured on a continuous scale. Instead, the claim would be that the lifetime lies within a certain range.

In water quality monitoring, the concept of no differences or trend needs to be carefully examined. There are differences, so the question becomes more one of how big are they? Thus when accepting a simple null hypothesis the real conclusion is that there is insufficient information to reject H_0 and to thereby accept H_A. It should be kept in mind that any trend or difference between groups, no matter how small, would be found to be statistically significant if the sample size were large enough.

With these concerns developing in the field of water quality data analysis, it is hoped that future texts make a clear distinction between the simple null hypothesis and the question of interest. It seems that the issue has received much more debate, if not resolution, in the behavioral sciences literature (Morrison and Henkel 1970, Oakes 1986).

It has already been noted that care must be taken when comparing "significant" results. The value of a hypothesis test result can be enhanced in two ways. The first is by also reporting the estimated confidence interval around the difference between the two sample means or around the magnitude of trend. Augmenting test results with estimates is advocated by Cochran and Cox (1957), and by Welkowitz et al. (1982). This removes some of the uncertainty posed by the "grey area" discussed above.

The second is by also reporting the "associated probability," or "*P*-value," of the test. This is the Type I error risk at which the test would just reject the null hypothesis. The *P*-value gives some indication of the strength of the test finding, rather than just a simple statement of whether the result was "significant" or not. The smaller the *P*-value, the stronger the conclusion of a difference between groups. Of course, *P*-values are directly comparable only for tests using the same sample size.

3.5.4. Goodness-of-Fit Tests

In early reports of water quality studies (such as Montgomery and Hart 1974) it was common to read that goodness-of-fit tests show that continuous water quality data are distributed as either normal or lognormal. These days the claim is not so common, and nonparametric methods are often advocated (Helsel 1987). A reason for avoiding characterizing data as coming from a particular distribution is that goodness-of-fit tests suffer from low power at small sample sizes. As records have extended, the tests have become more powerful. Also, in recent years tests more powerful than the traditional chi^2 test have been popularized. These would include the Lilliefors test (Iman and Conover, 1983:153), the related EDF Kolmogorov-Smirnov test (Zar, 1984:92), and the Shapiro-Wilk test (Gilbert, 1987).

3.5.5 General Implications For Network Design

One-sided hypotheses, as may be used to detect breaches of water quality standards, generally coincide with the question of interest. Two-sided hypotheses do not. However in the general case when simple null hypotheses are used (such an H_0 of no difference in means between two samples), it is not known a priori in which direction a change will occur. Therefore the two-sided test is more reflective of the real situation. In any case, hypothesis tests should be designed on the basis of power to detect a given difference. The authors discuss how to select sample size to achieve such a performance in Appendix A.

"Statistically significant" results may not be compared unless they are obtained from tests with the same sample size and significance level. As mentioned earlier, p-values should only be compared for tests using the same sample size. The design must recognize this when setting sample sizes.

Nonparametric tests are generally preferable over parametric tests. They are nearly as powerful as parametric tests when the assumptions of parametric tests are met. They are generally more powerful when those assumptions are not met. Their data requirements are much the same as for the equivalent parametric test, and so the data requirements of parametric methods can be used for nonparametric methods. It is perhaps well to keep in mind that nonparametric methods usually test hypotheses about percentiles, whereas parametric methods usually test more specific hypotheses on parameters such as means.

General texts on nonparametric methods are Bradley (1968), Hollander and Wolfe (1973) and Conover (1980). Conover and Iman (1981) and Iman and Conover (1983) point out that many parametric tests can be transformed into nonparametric tests merely by performing the usual parametric test on ranks of the data. A paired t-test becomes the Wilcoxon signed-ranks test for example. Gilbert (1987) presents many of the nonparametric tests that have been applied recently to water quality data. With the wide availability of computing power, the computationally-demanding randomization tests may become more widely used in the future (Conover 1980:327; Edgington 1986; Noreen 1989).

3.6 SUMMARY

In order to develop quantitative design criteria from the general questions that users want monitoring programs to answer, the following points should be kept in mind.

- The general questions are answered with uncertainty, using data obtained from samples. Statistical methods enable the uncertainty to be quantified and kept to a reasonable level.

- The general questions can be translated into specific design criteria, using the data requirements of appropriate statistical methods. These criteria typically include Type I and Type II error risks and magnitudes of differences to be estimated or tested. When there is agreement on these quantities, they become objective design criteria.
- Sampling regimes should ideally be strictly random. This is never possible, but the limitations should be realized. In particular there is a danger of missing a regular component of water quality variation when any time period has a zero chance of being sampled.
- Water quality data can be of different types and scales, and these dictate what methods may be used. Generally the best approach is to use nonparametric methods; their data requirements are similar to those for parametric methods.
- Hypothesis tests should be designed on the basis of power to detect a given difference. Comparisons of "statistically significant" results are valid only for identical sample size and significance level.

4

FRAMEWORK FOR DESIGNING WATER QUALITY INFORMATION SYSTEMS

When a water quality monitoring (information) system is defined to consist of sampling, laboratory analysis, data handling, data analysis, reporting, and information utilization, its design will necessarily have to incorporate a wide range of scientific and management concepts. From the chemistry of sample preservation through the statistics of data analysis to the incorporation of information into decision making, the design of water quality information systems is a large and difficult task. Throughout the task a focus should be maintained on generation of useful water quality information.

How does a person design a water quality information system that contains these six, very different but carefully related, components? Any approach to such design must recognize the magnitude of the design task. Topics as diverse as field sampling for biological organisms, analytical chemistry in the laboratory, data base management systems, statistically based data analysis packages, communications via reports, and utilization of information in decision making require a background that no one person will have. Consequently, design of water quality information systems will necessarily require a team approach with considerable time and effort devoted to the integration and documentation of all the components, as well as the actual design. In other words, while the *operation* of a water quality information system is mostly routine and hierarchical (through the six steps from sampling to information utilization), the design process is not. It calls for discussion among the information users, the operators, and the designers to arrive at a design that is feasible and appropriate.

An information oriented approach to design will have to be organized and systematic. A number of approaches have been described to help in the systematic design of water quality monitoring systems. Sanders et al. (1979) presented twelve steps to be followed in the design of water quality monitoring networks.

1. Determine monitoring objectives and relative importance of each;
2. Express objectives in statistical terms;
3. Determine budget available for monitoring and amount allocated for each objective;
4. Define the characteristics of the area in which the monitoring is to take place;
5. Determine water quality variables to be monitored;
6. Determine sampling station locations;
7. Determine sampling frequency;
8. Compromise previous objective design results with subjective considerations;
9. Develop operating plans and procedures to implement the network design;
10. Develop data and information reporting formats and procedures;
11. Develop feedback mechanisms to fine tune the network design; and
12. Prepare a network design report.

By 1983, Sanders et al. (1983) had reduced the number of steps to five by focusing each step on a major task to be accomplished before a water quality monitoring network design could be documented. The five steps are:

1. Evaluate information expectations;
2. Establish statistical design criteria;
3. Design monitoring network;
4. Develop operating plans and procedures; and
5. Develop information reporting procedures.

Mar et al. (1986) present four general tasks as a formal, systematic monitoring design procedure. These tasks are:

1. ''Identification of environmental changes of interest and the effects that would most likely manifest these changes,

2. Formulation of potential changes into conventional scientific hypotheses; definition of variables to be sampled, the technique to be used, and any alternative hypotheses that may cause the same type of change; and determination of whether complementary hypotheses provide the same information for decision making at lower cost,
3. Design of a cost-effective monitoring program to test each hypothesis by considering trade-offs between the improvement in statistical power (that is, the probability of discriminating between the various hypotheses) and the added cost of data acquisition, and
4. Integration of the individual monitoring programs by multiobjective ranking of all monitoring designs with respect to overall program goals."

Whitfield (1988) proposes a framework for designing water quality monitoring systems that includes five elements. On a site by site basis, he suggests the: (1) establishment of a monitoring goal; (2) selection of a sampling strategy to meet the goal; (3) periodic review of adequacy of sampling including quality control; (4) optimization of sampling related to the goal over time; and (5) review of adequacy of monitoring goal.

Schilperoort and Groot (1983), in discussing the optimization and design of water quality monitoring networks, also listed five steps which underscore the central role of defining objectives and the iterative nature of the optimization process. Their steps are as follows:

1. The monitoring objectives should be identified and quantified, including the definition of an effectiveness measure. This step, while being the most important step, is also a very difficult one. In general, these objectives include the estimation of the present quality state, the detection of long term trends, the detection of standards violations, and model studies.
2. The relevant processes should be identified, because their dynamics dictate the way the data should be analyzed. The dynamics depend on the water quality variables considered and the hydrodynamics of the water system.
3. The effectiveness of the information is determined by analyzing the data from the monitoring network. Often, the effectiveness can be related to statistical concepts such as the variance of the samples, the probability to detect changes or violations and the interpolation error.
4. The cost of the monitoring program is calculated.

5. The cost-effectiveness analysis is performed. Besides being difficult, this effort is often subjective. This analysis should, among others, take into account the relative importance of the sampling locations and variables. This type of analysis is almost always avoided in practice with only a minimum level of effectiveness specified a priori. Since the minimum level is based on subjective and/or political considerations, the real optimal sampling effort might not be achieved.

Hirsch et al. (1988), while not listing steps in the design process, discuss concepts involved in designing a national water quality assessment program. They discuss four strategic selections that are needed to make the assessment tractable. These are:

1. Selection of Scales (spatial, temporal, and focus);
2. Selection of Study Areas (study unit design);
3. Selection of Chemical Measurements (inorganic constituents, physical measurements, and organic compounds); and
4. Selection of Biological Measurements (fecal contamination, plant and animal tissues, ecological surveys, and biogeochemical processes).

They then discuss the use of available data, development of national data bases, quality assurance, communication and program coordination.

Shnider and Shapiro (1976) describe procedures for evaluating operations of water monitoring networks. In so doing, they indicate the contents of a monitoring system that should have been designed in the original establishment of the network. Their categories of activities to be evaluated, along with weighting factors for importance of each category, are as follows:

1. Network plan and design (0.167)
2. Personnel (0.152)
3. Facilities and equipment (0.141)
4. Sampling (0.150)
5. Quality assurance (0.157)
6. Data distribution and utilization (0.115)
7. Agency interactions (0.118)

The weighting factors were obtained from a survey of governmental experts in the field of monitoring. Shnider and Shapiro (1976) note that performing such an evaluation is very difficult and that to carefully evaluate all the complexities involved would be a lengthy and time-consuming process.

The General Accounting Office (1986), in evaluating five national efforts to assess water quality conditions in the United States, described five steps they felt are involved in any assessment of water quality. The five steps were then used to evaluate the monitoring system designs that supported each assessment. The five steps are:

1. Define water quality
2. Select measurable indicators
3. Collect data
4. Analyze data
5. Generalize from cases or samples to the nation

Smith et al. (1989), in describing the design of a national water quality network for New Zealand, list a number of tasks that require careful consideration if a successful network is to be designed and put into operation.

1. Develop the goal and objectives of the network;
2. Produce list of sampling sites (in detail);
3. Confirm statistical design criteria (including sampling frequency);
4. Produce a list of determinands (variables to be measured);
5. Assess accuracy requirements appropriate to the objectives;
6. Confirm laboratory analysis procedures;
7. Recommend appropriate laboratories;
8. Develop sampling routes;
9. Recommend field operators/groups and their training requirements;
10. Develop field sampling/analysis procedures;
11. Develop procedures for sample preservation and sample transport to laboratory;
12. Develop appropriate quality assurance procedures (i.e., from bottle washing, through analysis, to data entry);
13. Recommend data storage/retrieval/management system;
14. Recommend data analysis procedures and software;
15. Recommend reporting procedures (i.e., what, to whom, by whom, and when);
16. Assess cost of network.

Payne and Ford (1988) and Pollack and Ford (1989) present an overview of the design of a water quality monitoring program to assess the effects of changes in acidic deposition on surface water of the United States. The program, called TIME for Temporally Integrated Monitoring of Ecosystems, is to be a well planned extension of a number of ongoing monitoring efforts around the country. To set the stage for the design and operation of the new, coordinated monitoring program, a wheel-and-axle model was chosen to explain the overall concept. The wheels represent special surveys or intensive sampling efforts over space. The axle represents long-term, fixed-site, temporally intensive monitoring efforts focused more on answering questions about water quality changes over time.

Figures 4.1 through 4.4 indicate the wheel-and-axle model for four areas of the country where a variety of monitoring efforts are already underway (described by various acronyms), indicated by dashed lines on the axle. Where TIME begins, the axle becomes solid and coordinated with the wheels. The different figures contrast the variety of relationships between past wheel-and-axle monitoring efforts with the more coordinated relationships being designed under the TIME project.

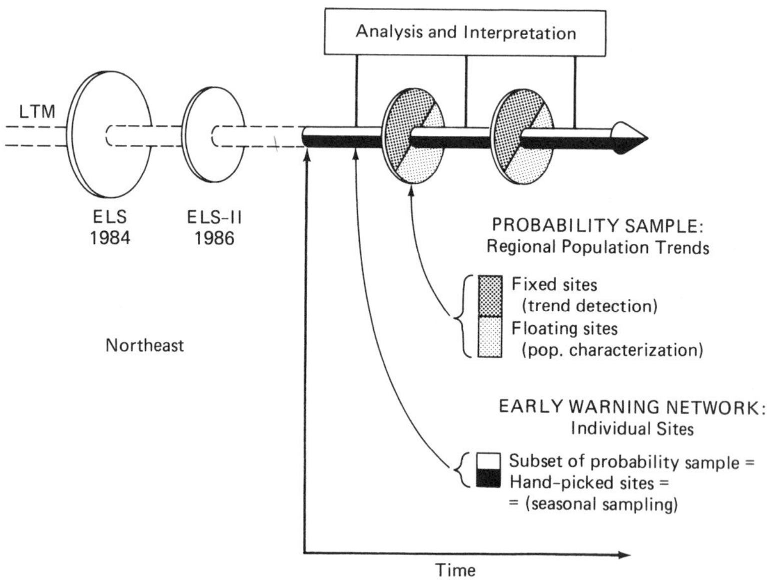

Figure 4.1 The conceptual "wheel and axle" design for the Northeast (from Payne and Ford, 1988).

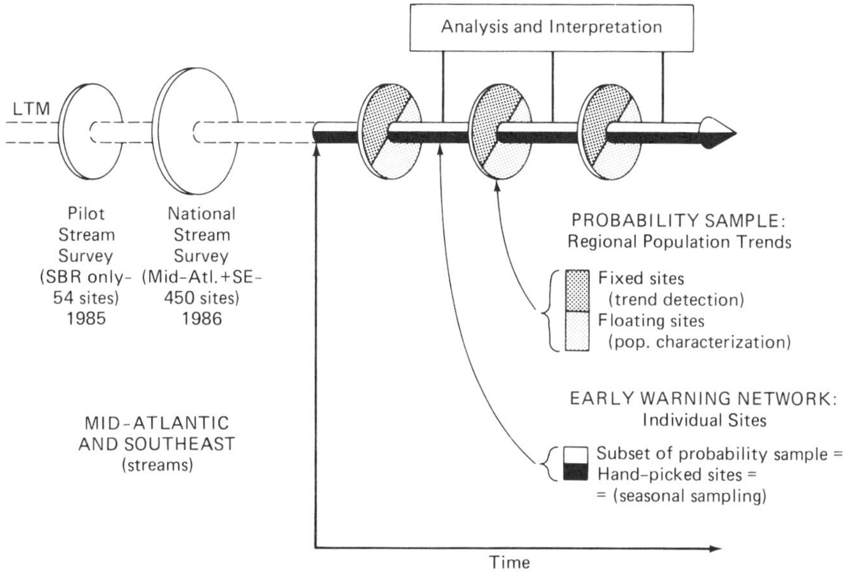

Figure 4.2 The conceptual "wheel and axle" design frame for the mid-Atlantic and Southeast (from Payne and Ford, 1988).

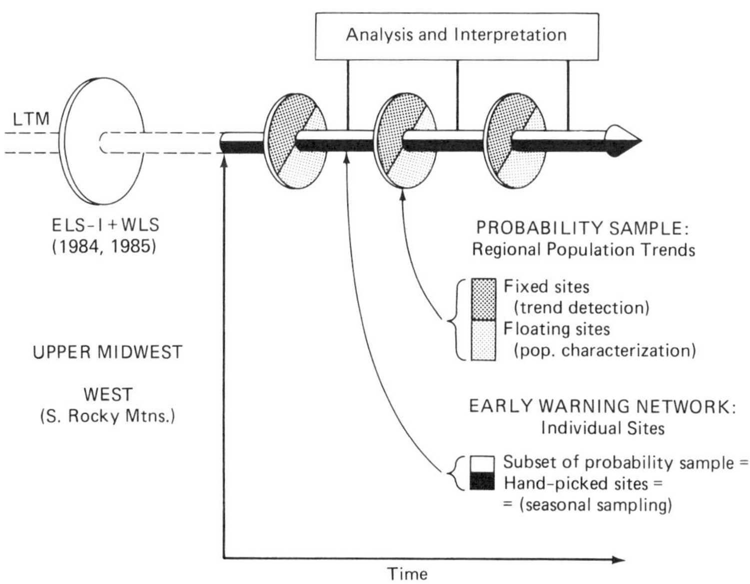

Figure 4.3 The conceptual "wheel and axle" design frame for the Upper Midwest and the Southern Rocky Mountains (from Payne and Ford, 1988).

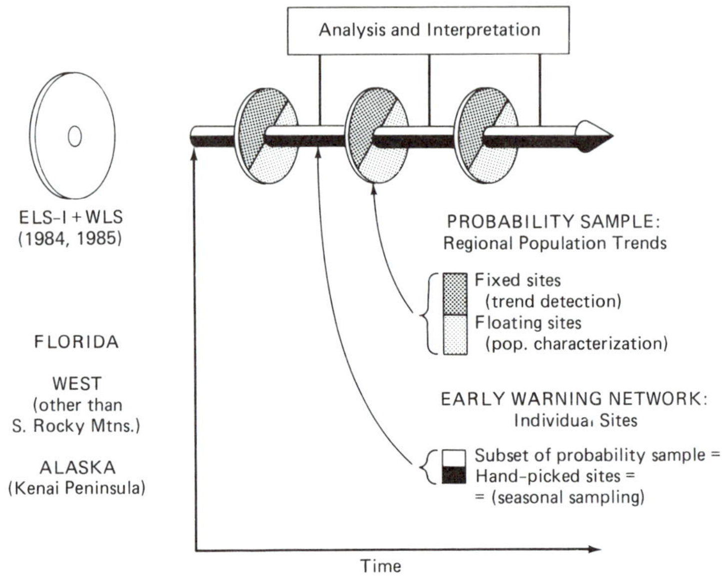

Figure 4.4 The conceptual "wheel and axle" design frame for Florida, in the West other than in the Southern Rocky Mountains, and Alaska (Kenai Peninsula) (from Payne and Ford, 1988).

These many efforts to describe how one might logically approach the design of a water quality monitoring program have many common attributes. In the following section, a design framework will be presented that builds upon these commonalities.

4.1 DESIGN FRAMEWORK

Ward (1986) and Ward et al. (1986) further elaborated on the five steps (from Sanders et al. 1983) to the design of a water quality monitoring network by focusing more heavily on the need to establish information goals for monitoring and to quantify information objectives in a way that relates to the graphical and statistical results of data analysis. This increasing focus on quantifying information objectives and relating them to the details of the monitoring system design is also used in this book.

A framework for design which focuses on information produced must have design steps that address the information needs of management and the ability of the measurement system to supply information. Thus, a first step in the framework will be to quantify the use of water quality information within the management effort. Information goals can then be established in view of these intended uses.

The next question relates to the ability of a measurement system to produce the desired information. It is one thing to say that management needs such and such information on water quality, but it is quite another to actually produce the information via a monitoring program. In fact, this is perhaps one of the major failings of past monitoring design practices—the information expectations placed on monitoring have been far greater than the capability of monitoring to produce. These expectations have rarely been quantified, and often the designer does not find out about them until the first reports are prepared. At this time a manager may express dissatisfaction with the information because it does not meet his/her expectations. These expectations, however, were never quantified and used as part of the monitoring (information) system design process. *And* the designer never quantified the information limitations of the monitoring system so that the information users could formulate realistic information expectations.

Another element of past approaches to monitoring, that relates to why the above situation developed, deals with the heavy focus on legal approaches to water quality management, especially in the U.S. When monitoring is legally required, it is often conducted for the legal reasons alone, and the scientific or management information needs are ignored. This often stymies questioning of the relevance of monitoring to the information needs of an agency. The monitoring is required by law and, thus, must not be questioned. This situation is the reverse of the problems of management's unrealistic expectations for a monitoring program in that legal requirements often lead to monitoring programs with only vague information expectations. In the recommended design approach, one does not necessarily question the need for monitoring but does require that the information produced be relevant to management decision making.

Thus, the framework for design of water quality information systems begins with quantifying the information required by management and quantifying the information that the monitoring system is capable of producing. As this statement implies, there can be differences between the information desired and the information which can be produced. This difference, if it exists, must be resolved at the beginning of the design process. Failure to resolve such differences early places the entire design process on a very fragile basis and will only lead to frustration on the part of the information users and operators of the monitoring system. It also leads to a high probability of the monitoring system being deemed a failure several years into the future.

As discussed earlier, monitoring provides the information demanded by the water quality management strategy implemented from the law. Interestingly, however, the management strategy chosen to implement a water quality law may have to be modified to reflect the ability to supply the information needed to administer the law accurately and fairly. Thus, not only should one question the design of the water quality monitoring system but also the management structure that may be demanding more information than the monitoring system is capable of producing. This is especially true in today's economic environment where monitoring resources are often reduced to help balance a water quality management budget.

Given the above, the first step of the design process will be to quantify, as far as is possible, the information required by management. The second step, but not necessarily in sequence, is to quantify the information the monitoring system is capable of producing. If after completion of the first two steps, there are differences in the information expectations of management and the ability of monitoring to produce information, then the management strategy, the monitoring budget, the monitoring system design, and the law itself may have to be examined and reformulated. This may seem over-stated; however, it is simply a waste of money to monitor without a clear relationship between the information to be produced and its use within the management agency's decision making process! It is the purpose of the water quality information system design to ensure that monitoring is well integrated into the information needs of water quality management.

It must be noted that information produced by the program will be used for purposes other than those identified in the initial design. That is fortuitous. It is, however, not a reason to collect data in the vague hope that someone may eventually find a use for it. Sensible design calls for quantifying information needs, so far as they may be known.

If the monitoring system being designed is only a part of the total effort to obtain information on water quality conditions for a management program, its relationship to the other monitoring efforts must be defined. The wheel-and-axle model, as described by Payne and Ford (1988) and Pollack and Ford (1989) presents a means of indicating connections between various monitoring efforts. Figure 4.5 is an example of a generalized wheel-and-axle model that can be utilized to describe a total water quality information system consisting of several separate, but related, monitoring efforts. This graphical visualization of relationships between

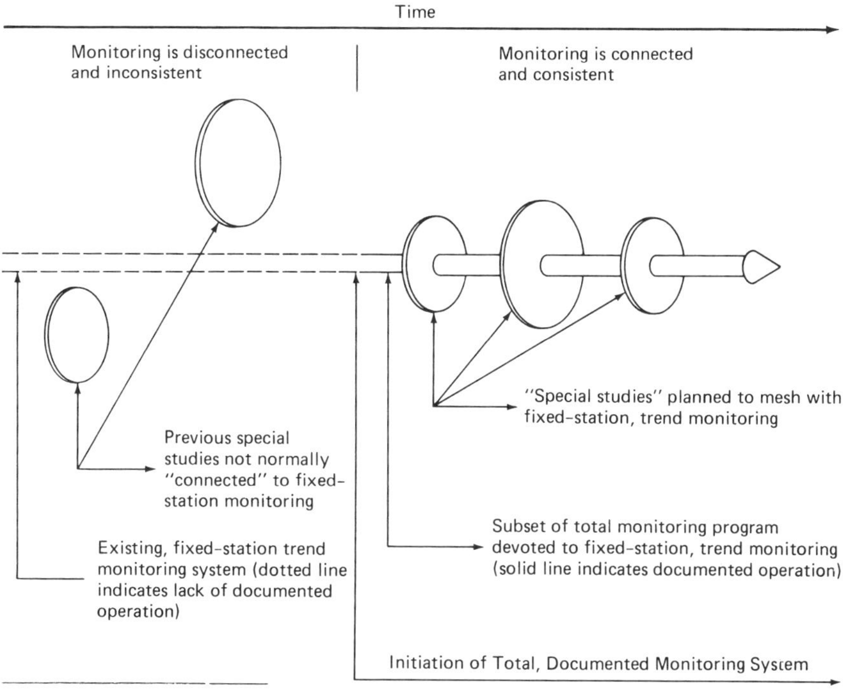

Figure 4.5 Generalized "wheel and axle" framework for developing a fully integrated water quality monitoring system.

various monitoring efforts can be very useful in communicating the coordination to be achieved by a more systematic approach to monitoring system design.

Once the information required by management is balanced against the ability of monitoring to produce information, the next step is to actually determine where to sample, what to measure, and how frequently to sample. With the information needs quantified, determining these aspects of design will be straightforward. These three determinations, for many years, defined water quality monitoring system design. However, without a clear definition of the information to be produced, their determination was much more difficult. The National Academy of Sciences (1977), perhaps better than anyone else, summarizes this issue in the following quote:

> "In recent years there has been a number of research reports, some published by the EPA, to describe various methods for designing ambient water quality monitoring systems (Ward, 1973; Sanders, 1974;

> . . . .). They fail to demonstrate that users and operators of networks agree on criteria that are relevant to their needs. Because EPA did not supply the initial criteria for design, many of these reports differ in their techniques and fault the assumptions and criteria developed in other reports."

While the U.S. EPA (and the first author of this book, Robert C. Ward) is being criticized in this case, almost any agency involved with water quality monitoring could have been criticized as well. There were too many "network" designers in the 1970s who attempted to establish their own design criteria for the "where, what and when" of monitoring, using their own opinions of what information was needed. There was little attempt to actually quantify the information needed to operate a management agency and then design the monitoring system to supply this information. This is the point the National Academy of Sciences (1977) is making above.

With the information needs defined, however, it is much more straightforward to determine the "where, what, and when" of water quality monitoring (that is the network design) and be assured that useful information will be produced.

Once it is known where samples will be collected, what will be measured, and how often the samples will be collected, the next step in the design framework is to define the exact procedures to be followed in operating the entire monitoring (information) system per the definition provided in Figure 1.1. Poor definition of many, if not all, of the components of monitoring systems may result in variations in the monitoring system procedures which, in turn, adds to the variation seen in the water quality data. Inconsistency in operation often means that the data analysis component of the monitoring system is actually analyzing variation in the monitoring system itself and not variation in water quality in the environment. To build into the monitoring system the consistency necessary to ensure that the variation in the data is due to water quality, requires complete specification performance standards for all details of the system's operation. That is, the design is an exercise in *quality assurance*.

Thus the fourth step in the design framework requires documentation of all procedures involved with collecting samples, laboratory analysis of the samples, data handling, data analysis, reporting, and information utilization. This is a huge documentation effort that is eased by the fact that some documentation already exists as standard methods or protocols. There is a danger that if the documentation of data collection is blended with documentation of the information generation portion of the system, the latter portion may be slighted due to its poor definition in the literature. For this reason the design framework suggested in this text will include a fifth step that, in effect, splits the fourth step into two parts. Step 4 focuses on data collection, and step 5 focuses on documenting the information generation activities as they relate to the information needs defined in the first step of the framework.

Reporting water quality information often requires a number of different types of reports addressing a hierarchy of information needs: the public, policy makers, administrators, and technical staff. The frequency of reporting will vary as technical staff need more frequent information than, say, the top level policy makers. Reports must not be simply tables upon tables of numbers. There should be a designed method for converting the numbers into the desired information for each audience. Lists of data, if they are included, should appear in the appendix or in separate reports designated as data only. The many ways one can present information from data should be made use of. Tufte (1983) is an excellent discussion of ways to visually display quantitative information.

The entire framework for designing a water quality information system, as presented above, can be summarized in the following:

Step 1. Define information needs of management

- identify information needs of each management tool
- summarize information needs of agency
- relate agency information needs to monitoring strategy
- define reporting and information utilization procedures desired by management
- determine appropriate statistical means for producing the desired information

Step 2. Define information that can be produced by monitoring

- statistically characterize water quality "population" to be sampled
- review statistical methods applicable for generating the desired information, including their data requirements
- state what information can be produced
- compare information sought with information that can be produced

Step 3. Design monitoring network

- document sampling locations
- determine what to measure
- compute sampling frequency

Step 4. Document data collection procedures

- field sampling operations and procedures
- laboratory analysis methods and operations
- data storage and retrieval system

Step 5. Document information generating and reporting procedures

- data analysis hardware and software
- reporting formats and frequency
- information utilization procedures

4.2 PLANNING FOR THE DESIGN OF A WATER QUALITY INFORMATION SYSTEM

As noted earlier, the above framework for designing a water quality information system involves a considerable amount of work and time. The ad hoc design methodologies of the past, which required little time and effort, are often on the minds of a manager when a water quality monitoring system is to be designed or redesigned. The expectation is that the design or redesign can be done by existing staff within several weeks or months while they continue their regular duties. Hopefully, the previous discussion has revealed the extensive effort required to create a true information system quantitatively supporting water quality management.

In planning for the design or redesign of a water quality monitoring system (in hopes of creating an information system), the time and effort required must be recognized by those managing the agency. The resources needed for a thorough design effort must be quantified and budgeted. The ultimate, potential impact of such a thorough information system design on the management agency's operations must be understood and accepted as desired output of the design process. Managers must understand that in order to design a truly supportive information system, their use of information will be questioned and, ultimately, quantified. This does not mean that their decision making authority will be usurped. Managers will still be left to make decisions which must incorporate a wide variety of information (economic, political, social as well as environmental). The designer of the information system needs a quantitative definition of the water quality information that managers want to incorporate into their decision making process.

5

QUANTIFYING INFORMATION EXPECTATIONS

One of the most important and difficult tasks in information system design is writing down exactly what information is to be produced by the monitoring effort (or is to be expected by the information users). To begin, the definition of "information" should be agreed upon. Information might be defined as "knowledge gained in any manner including facts and data." However, in an information age this definition should be amended to read *useful* information gained from the organization, analysis, and reporting of data for a *particular purpose*. This definition is particularly relevant in water quality monitoring, where for too many years data were collected without a data analysis plan being developed before sampling was begun.

How does one go about writing down the information to be produced from a water quality monitoring effort? Logically one might first consider the information needs of management and perhaps identify those needs as the information expectations of monitoring. However, as the authors indicated in Chapter 4, the information needs of management and the information which a realistic monitoring system can actually produce are not necessarily the same.

Normally developing useful and attainable information goals will require iteration between the needs for information and the ability of a monitoring system to satisfy those needs. Perhaps the first step in quantifying information expectations should, therefore, be to define the types of information that monitoring can produce.

5.1 TYPES OF WATER QUALITY INFORMATION

By identifying the information types which can be produced, it is less likely that naive and unrealistic information expectations will be placed on monitoring. The following classification of information types is organized around the methods by which information is obtained from data and conveyed to users.

5.1.1 Narrative Information

Narrative information is particularly useful for communicating with the lay public. The primary difficulty with this type of information is that it can be imprecise. Also wording can be chosen to convey a particular desired impression on the state of water quality which may not agree with factual data. For example the statement ''Water quality conditions in Colorado are generally good'' conveys little information but will not raise any red flags either. The statement ''Stream standards are met on 94% of the stream miles in Colorado'' creates an impression of good water quality while the converse ''Stream standards are not met on 6% of the stream miles in Colorado'' may create a different impression.

More factual statements such as ''The water surface is free from visible oil and grease'' are less subject to misinterpretation and may, in fact, be sufficient to indicate compliance with a narrative standard. Thus, in writing down information expectations, it is crucial that the wording used in stating the results of monitoring relative to narrative standards be defined in the initial design of the information system.

5.1.2 Numerical Information

Water quality observations (raw data) themselves are a form of information that is perhaps more readily identified with past monitoring results. Many of us have seen monitoring reports that are nothing more than table after table of the raw data listed by site and date. The presentation of the results of monitoring in this form generally implies that the data are the end product of the monitoring system. The user is to read the data and thereby obtain the information needed. Those who collect the data are simply supplying numbers with no interpretation, summarization, or analysis.

In fact there are times, particularly when monitoring is part of a regulatory action, that monitoring system managers are told not to analyze or interpret the data for fear that the interpretation will be used against those performing the data collection. The expectation of monitoring, in such a situation, is to simply meet the letter of the law. Under such circumstances, the monitoring system designers and managers have as their information goal the production of numbers!

Admittedly, production of numbers may sometimes be sufficient. This often occurs in the situation where the operating limits of a wastewater treatment plant, for example, are defined as set numbers not to be exceeded. Measurements of the effluent are taken and compared to the limits. The use of the monitoring results is achieved when the comparison is made. The monitoring system, in this case, is designed to operate with a well defined information purpose and data analysis plan—comparison to set limits. Of course, if such comparisons are an information goal, then the report should contain the comparisons.

A problem may arise, however, when someone later comes along and wants to know whether there has been a trend in the quality of water discharged from the plant. The data were not collected for this purpose, and no data analysis plan has been established to achieve this information goal. The data can still be analyzed for such trends, but those wanting the information should not be surprised when the trend analysis is less than satisfactory due to problems with the data not being directly compatible with the data analysis method chosen.

Raw data can be an end information objective or, as is more generally the case, a step toward graphical and/or statistical interpretation to achieve a specific information goal. The design of any water quality information system must specify how the raw data are to be used.

5.1.3 Graphical Information

Graphical displays are probably the most useful approach all around for conveying information to a wide variety of audiences, both technical and nontechnical. In general, graphical displays are easier to construct, more interesting, and less likely to be misunderstood than narrative descriptions. Quantitative as well as qualitative information can be portrayed graphically. For example, box-and-whisker plots (Tukey, 1977; and Ward et al. 1988), give a visual indication of data distributions (central tendency and spread) and indicate how distributions change over time or between locations.

Box-and-whisker plots are very simple to construct. The box contains the middle 50% of the data (interquartile range) with the median indicated by a horizontal line. The whiskers describe the more extreme conditions or tails of the distribution. The whiskers may extend to the minimum and maximum of the recorded values or could extend only to specified percentiles, say the 5th and 95th, with outlying values plotted as dots beyond the whiskers. There are many variations on the general format for box-and-whisker plots (Tukey, 1977; McGill et al. 1978; Velleman and Hoaglin 1981; and Tufte, 1983). Box-and-whisker plots are especially effective in comparing data records.

A seasonal box-and-whisker plot, such as that shown in Figure 5.1 portrays seasonal changes in both the central tendency and spread of the data. In the example, the data are grouped according to quarters, each group having its own box-and-whiskers. In Figure 5.2, the same data are grouped according to year. This annual box-and-whiskers plot shows how the distribution is changing from year to year and may be more readily interpreted than the time series plot of the same data, shown in Figure 5.3.

Graphical methods are also useful as a check on statistical methods. For example one should not generally believe that a time trend in water quality is "significant" if it is not apparent from a time series plot or an annual box-and-whiskers plot of the data, unless, perhaps, the data record is extremely long.

The designer of a water quality information system can formulate and document the graphical means that are to be a part of the information

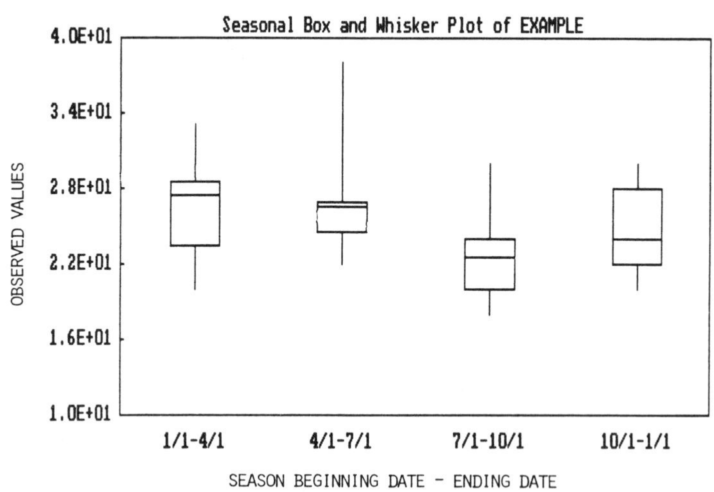

Figure 5.1 Seasonal box-and-whisker plot of example.

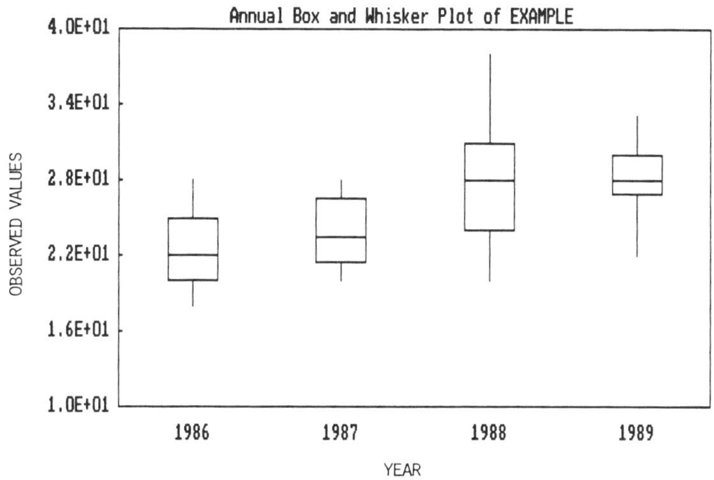

Figure 5.2 Annual box-and-whisker plot of example.

reporting to take place as data are collected in the future. The documentation should clearly state and illustrate the type of graphics to be used, how they are to be developed, and what type of information is expected to be obtained from each type of graph.

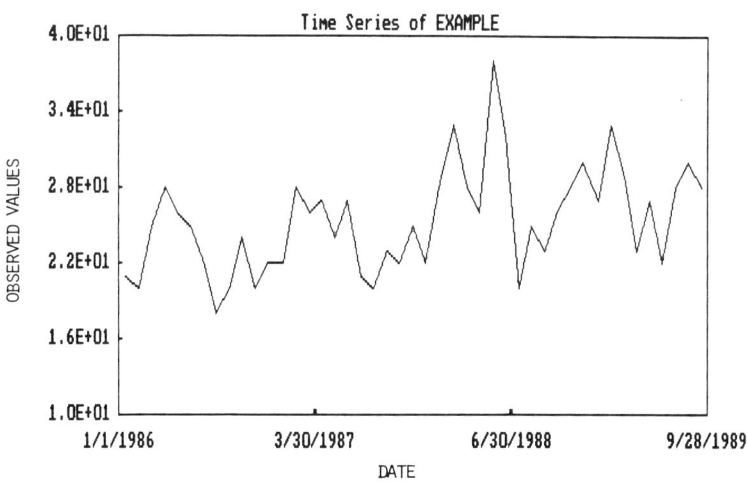

Figure 5.3 Time series of example.

5.1.4 Statistical Information

As was discussed in Chapter 3, statistical information is an extremely important part of any effort to use a few samples from a "population" to make inferences about the entire population.

Statistical methods permit one to make quantitative statements about natural processes which involve error and uncertainty. Absolute statements about water quality such as "The stream standard of 10 mg/L is being met 100% of the time" are rarely appropriate and rarely meaningful, given the uncertainty associated with monitoring. More meaningful statements would include expressions of the level of uncertainty resulting from the sample size used in the analysis and the variability of the observations.

Statistical information is, therefore, the most useful type for making quantitative decisions, such as whether water quality is improving or getting worse over time, whether a particular remediation effort is effective, or whether a particular standard is being met. The latter, of course, assumes that there is a definition of "standards compliance" that accounts for uncertainty, both in water quality observations and in inferences drawn from those observations regarding water quality conditions.

Information produced from a water quality monitoring system can provide an understanding of the following:

1. Average water quality conditions;
2. Changing water quality conditions; and
3. Extreme water quality conditions.

The statistical methods for achieving this understanding can involve a number of approaches.

Statistical information falls generally into one of two categories: *summary statistics* (sometimes portrayed graphically), such as mean or median and standard deviation or interquartile range, and *tests of hypotheses,* such as the classic "*t*" test for difference in means. Estimates of population parameters, such as means, are generally most useful when expressed as intervals (giving an indication of the level of uncertainty associated with the estimate) rather than as points or single values. The authors have discussed estimation and testing of statistical hypotheses at some length in Chapter 3. From an environmental monitoring perspective, interval estimation for means, variances, quantiles, and proportions is well covered in Chapters 11 and 13 of Gilbert (1987).

A formal test of the null hypothesis that no trend exists in our earlier example data from Figures 5.1 through 5.3 can be performed using the seasonal Kendall test. The resulting test statistic is 2.911 (standard normally distributed), which would cause rejection of the null hypothesis at significance levels greater than 1%. Figure 5.4 shows the time series with a superimposed trend line with slope of 2.0 units per year estimated with the seasonal Kendall slope estimator (Hirsch et al. 1982 or Gilbert, 1987). Although this trend was fairly obvious from both Figures 5.2 and 5.3, the formal test with a specified procedure provides a means for any data analyst to reach the same conclusion regarding whether or not the trend is "statistically significant."

A discussion of specific statistical methods useful in the field of water quality data analysis will be presented in Chapter 6 along with a framework—a data analysis protocol—to organize the use of statistics.

5.1.5 Water Quality Indices

A water quality index combines information from several water quality variables into a single number. This number may reflect the suitability of water quality for a specific purpose. Water quality indices have been used less frequently than air quality indices, possibly because of the difficulty in reaching agreement on how water quality should be defined. An exception would be biological indices associated with special surveys.

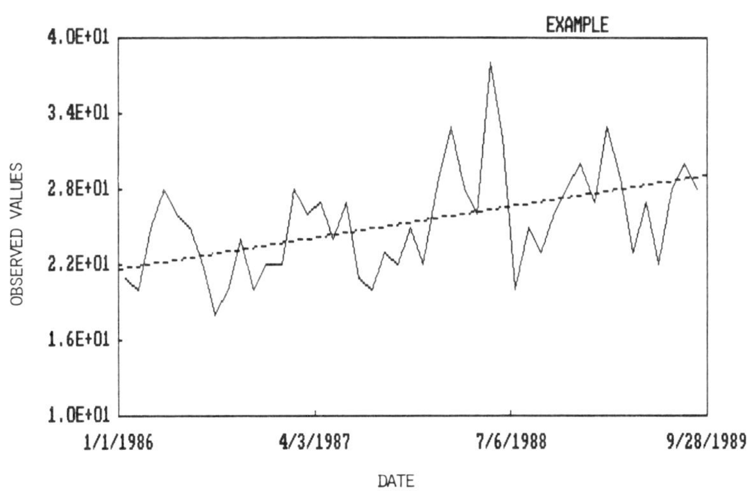

Figure 5.4 Time series of example with superimposed trend line.

However, the concept of indices remains promising for conveying the results of routine water monitoring to non-technical users and to the public.

There were a number of efforts in the 1970s to study and develop water quality indices for use in water quality management. Brown et al. (1970) argue that water quality indices are needed in water quality management. Landwehr (1974 and 1979) evaluated the various methods being used to construct water quality indices and their statistical behavior. Dunnette (1979a) presented the development of an index being used in Oregon and then (Dunnette, 1979b) showed how such an index could be used to optimize sampling frequency. Ball and Church (1980) review the development of water quality indices and summarize several of the mathematical formulations being proposed. Ott (1978) provides an excellent discussion of the subject and is still the "standard" reference.

During the 1980s the focus on water quality indices turned toward application. This is not to say that the theory is complete, but there was a notable absence of theoretical development taking place during the decade. Perry, Ward, and Loftis (1983) surveyed state water quality management programs' use of water quality indices. Twenty states were using some form of an index to report on the status of their water quality. Smith (1989) describes the development and proposed application of four water quality indices for New Zealand's rivers and streams.

5.2 LEGAL BASIS FOR MONITORING AND LEGAL REQUIREMENTS FOR INFORMATION

With an understanding of what information can be produced from a water quality monitoring system, the authors examine what information is required for water quality management. Ultimately, the goal of a design effort is to match the two. Information needed for management decision making is to match information produced from the monitoring system.

In determining what information is needed to support water quality management, it is often necessary to go to the very foundations of the entire water quality management effort. Those currently involved in management activities may sometimes express a desire for information which is not founded on legal or scientific rationale. Since the majority of water quality management efforts are the result of legislation to control water quality for maximum benefit of society, one may go to the laws to begin the search for exactly what information is needed to manage water quality. This exercise can be very enlightening, even though, as noted in Chaper 1, the laws are generally not very clear on the subject of information requirements.

In examining laws establishing water quality management efforts, one sees three general categories of information demands. These categories, as noted in Chapter 1, are the following:

1. Information needs implied by the general water quality goals of the law;
2. Information needs implied by management's functions and powers; and
3. Information directly required by reporting/monitoring provisions.

5.2.1 Monitoring Implications of Legal Goals

After a water quality law is passed and then implemented, the question is eventually asked, "Are we achieving the water quality goal(s)?" Reilly, (1989), and the Committee on Science and Technology, (1983)—hearings on national environmental monitoring—provide a feel for the questions the lawmakers eventually ask of management and its monitoring effort. If the monitoring system has not been designed to answer these questions, it is unlikely that a satisfactory answer will be abstracted from existing water quality data—although several gallant efforts have been undertaken (Wolman, 1971; Smith et al. 1987; and the five efforts evaluated by the General Accounting Office, 1986). Whose responsibility is it to anticipate the questions about water quality goal attainment and to design monitoring systems to provide this information?

Examining the water quality goal statements that have been included in water quality laws reveals the difficulty in developing a monitoring system design which will be able to provide the implied information. For example, the following goal statements contain quite different implications for measuring their attainment.

1. "to promote the conservation and best use of natural water"
2. "to maintain or improve the quality of natural water"
3. "to prevent, abate and control water pollution"
4. "providing for the protection and propagation of fish, shellfish, and wildlife, and for recreation in and on the water . . . wherever attainable."

Thus, if measuring goal attainment is a purpose for water quality monitoring (and it certainly is implied in water quality laws), then it is critical that a clear water quality goal be formulated for monitoring system design purposes. Many water quality laws have numerous "goals" that may be conflicting when it comes to measuring attainment.

To further expand, consider the current U.S. water quality "working" goal of "fishable and swimmable." To assess attainment of the goal, what information needs to be obtained? It appears that information on fish and swimmers is needed. What fish are to be present for a stream to be classified as fishable? How are the fish to be sampled? Where are they to be sampled? How much confidence is needed in the statistical result? Is it necessary to sample every stream reach? Will one sample per year be sufficient to state that a stream reach is fishable? The list of questions goes on and on! However, it is this type of questioning of legal water quality goals that must precede any effort to design a monitoring system to measure attainment of such goals. For example, Smith et al. (1989), in establishing a national water quality monitoring network for New Zealand, related the monitoring goals and objectives to sections of the laws that established the organization performing the monitoring activity. The U.S. Environmental Protection Agency (1974) reviewed its statutory authority regarding monitoring relative to its implementation responsibility.

5.2.2 Monitoring Implications of Management Functions and Powers

As noted in Chapter 1, water quality management is provided with various means to carry out the provisions of the law. There are criteria, standards, permits, enforcement procedures, construction loans, planning requirements, and research functions and powers that can appear in a water quality law. Each of these management "tools," to be implemented effectively, requires information on the behavior of water quality in the environment. The exact nature of the information needed to implement each function or power is usually not specified in the law. Considerable latitude exists for the water quality management staff to influence the information demanded and the information used within the management effort. Also, at this point, it might be helpful to examine specific regulations that have been developed to help with implementation of the legal functions and powers. Many state and Federal regulations are beginning to include explicit needs for monitoring. RCRA, being perhaps the most explicit, gives guidelines on what to monitor, how often to sample, how to analyze the data, and what to do when certain statistical conclusions are reached (U.S. Environmental Protection Agency, 1986a and 1989). In monitoring associated with hazardous waste management, in particular, data and information requirements are often dictated by a need to convince a judge or have the information stand up in court. These requirements, unfortunately, may be far in excess of, or different from, those needed to gain an understanding of water quality for management purposes.

The efforts of society to address water quality problems have resulted in separate laws addressing a specific problem or specific portion of the hydrologic cycle. This fragmented legal framework has created a fragmented management effort which various state and Federal agencies try to rationalize into a logical and efficient structure. As one might expect, fragmented legal requirements generate fragmented regulatory information demands and specific monitoring systems for each regulatory activity. It is not uncommon to find a dozen separate water quality monitoring programs within a state water quality management effort, each addressing a specific regulatory information request.

When examining the information implications of the functions and powers of water quality management efforts, it is important to state clearly the information to be supplied by a monitoring system and to which function and/or power it relates. Ideally, all monitoring efforts should be coordinated into one water quality information system supplying the overall information needs at the same time specific regulatory needs are being met. The "wheel and axle" concept, described in Chapter 4 for relating various monitoring efforts to one another, is one such coordinating strategy.

5.2.3 Directly Stated Information/Monitoring Requirements

There are often statements in water quality laws that call for water quality assessment and/or monitoring. Phrases such as the following may appear, "The Division shall take such samplings as may be necessary to enable it to determine the quality of every reasonably accessible segment of state waters." These statements do not define the nature of the information to be produced nor its use in management decision making. They are catch-all phrases that permit monitoring to be undertaken and are not helpful to the monitoring system designer in defining information goals.

Reporting is often specified in water quality laws. Section 305(b) in the U.S. Clean Water Act requires national assessments of water quality every two years. The exact procedures for the assessment are not defined in the law but left to management staff to determine. The U.S. Environmental Protection Agency (1985a) has been developing and refining procedures for conducting such assessments for a number of years.

Thus, even when water quality monitoring or reporting are directly stated in a law, there is very little guidance for the monitoring system designer. Considerable investigation of information objectives, as described in Chapter 4, must be undertaken.

5.3 MANAGEMENT STRUCTURES AND STRATEGIES AND THEIR INFORMATION NEEDS

The management structure and strategies of a given organization will strongly influence information needs and expectations from monitoring. In examining the various ways that a management agency's powers and functions are implemented, a monitoring system designer has to carefully assess the regulations and guidance documents being prepared as a part of the implementation strategy. The following list presents the type of questions that a designer may ask relative to the implementation of management's functions and powers:

1. In establishing water quality *criteria* applicable to a given use, how is past water quality data utilized?
2. In *classifying* a water body for a given use, how is water quality information from that water body used?
3. In performing a wasteload allocation for issuing a *permit,* what water quality information from the receiving water body is needed?
4. In measuring *compliance,* what definition of compliance is used, and how does it relate to water quality data/information obtained from the receiving water body?
5. If *enforcement* actions are necessary, what water quality information is needed for each of the steps in the enforcement process?
6. When a construction *grant or loan* is evaluated, what water quality data/information is used in the decision making process?
7. When a *planning* effort is undertaken, what water quality information is required or desired?
8. In *assessing* achievement of water quality management goals for water quality in the environment, what water quality data/information is required?

Examining the various powers and functions in such an inquisitive manner will help define the information needs of management. There are several important factors to consider:

1. Who are the decision makers with regard to implementation of each power and function?
2. How are decisions made? Are data or results of data analysis used directly and applied in set rules, or are decisions made "by the seat of the pants"?
3. How timely must water quality information be to effectively support decision making?
4. To what types and what amount of information will management respond most favorably. Some managers expect very brief reports which get right to the point. Others are impressed by long reports and expect to have every detail included, even if the reports are never read. Almost everyone likes fancy graphics, preferably in color. In some cases, information can be conveyed routinely to management via computer, with emphasis on graphics and no paper involved.

To illustrate, consider a situation where a management program is to assess success in achieving a stream water quality objective. The objective may be to "maintain or improve" water quality, as measured against a standard. Water quality standards are developed for streams in segments, depending upon use of the water. Is the strategy for assessing achievement of the objective to measure each segment with one or two samples per year, or is it to identify representative segments and obtain enough samples to develop statistical "confidence" in the monitoring results at each site? Will the information be spatially intense with low confidence in each result, or will it be temporally intense with higher confidence but less spatial coverage? The answer to these questions often comes from a management structure that does not appreciate the application of statistics to environmental assessment. Either strategy requires careful application of statistics to the design of the monitoring effort.

5.4 PUTTING INFORMATION REQUIREMENTS INTO WORDS

After the legal and regulatory review of the information required for management purposes, the designer should write down his/her assessment of these needs, along with an evaluation of what information can reasonably be produced, and use this evaluation to initiate a dialogue with information users. Such a written report can take considerable effort, but it can initiate an enlightening discussion of management strategy versus monitoring/information needs and resources.

In preparing a report on management strategy, information needs, and monitoring limitations, a monitoring system designer may be placed in an uncomfortable position. The designer is questioning how people are doing their job and questioning how managers or policy makers make decisions. Such a task should only be undertaken with complete support of the upper level managers in the management program, or it is doomed to fail.

To illustrate, Loftis, Porter and Settembre (1987) describe an effort to define information expectations within an industrial water quality management effort. They found that perhaps the most critical part of formulating information needs is gaining input from data and information users via formal or informal meetings, questionnaires, or interviews. They found this to be a very frustrating task.

When asked what information is needed, users often reply "Everything, I need all the data." Even if users cannot tell you how the data are used, the answer probably will not change. Loftis, Porter, and Settembre (1987) found the legal assessment of information needs, along with a sample report of expected water quality information, to be useful in indicating to the users how the monitoring system designers, saw the situation. By letting the information users react to the information expectations report, the water quality personnel (information users) could indicate what they would like to see changed.

In the industrial example, the management staff was involved in treatment control at a supervisory level, making daily, weekly, and monthly decisions which were supported by extensive monitoring. Real-time or minute-by-minute control was considered as a separate function and not directly included in the study.

From one-on-one interviews with staff, written questionnaires, and numerous group meetings, seven specific management objectives were identified, each of which required information from monitoring:

1. Evaluating whether the treatment process was performing satisfactorily
2. Detecting treatment upsets such as slugs of toxics and/or die-off of treatment microorganisms.

3. Backtracking slugs or spills from the treatment plant to their source
4. Meeting state discharge permit requirements
5. Evaluating impacts of plant discharges on receiving waters
6. Evaluating trends in treatment loads
7. Evaluating trends in concentrations of constituents at key locations in the waste stream

While the above list of information objectives was regarded as a significant step forward in achieving an effective information system, it is appropriate to note that complete agreement on the list was never achieved.

5.4.1 The Information Expectations Report

In preparing an information expectations report (i.e., putting the information expectations into words), it is useful to first describe the setting within which the water quality management effort is taking place—the geography, hydrology, water quality problems, laws established to deal with the problems, along with a description of the management program. Next, an analysis of the information needed by the management structure and strategy currently in place should be included along with an assessment of the monitoring ability to produce such information. A summary of the tasks required to prepare and present an information expectations report is contained in the following:

I. Perform a thorough review of the legal basis for the management effort and define the resulting ''implications'' to monitoring
   - Information expectations implied by legal objectives
   - Information expectations implied by management's functions and powers
   - Directly stated monitoring requirements

II. Review administrative structure and procedures developed from the law in order to define information expectations of management staff
   - Standard setting procedures
   - Wasteload allocation/permit procedures
   - Enforcement procedures
   - Planning methods
   - Water quality assessment methods
   - Current utilization of monitoring results for management decision making

III. Review ability of monitoring to supply water quality information
    - Describe alternative monitoring approaches available
    - Define information each is capable of supplying (narrative, numerical, graphical, and statistical)
IV. Formulate information expectation report for monitoring system
V. Present information expectation report to all users of the water quality information
VI. Develop consensus as to an agreeable formulation of information expectations and related monitoring system design criteria (have those agreeing to the information expectations sign the report)

To further elaborate, the following contains the outline of what an information expectations report might look like.

1. Evolution of Water Quality Management Program
    - Geographical/hydrological setting
    - Water Quality Problems
    - Water Quality Laws
    - Management Program Structure
    - Management Procedures
2. Information "Expected" by Management Program
    - Implications of the Law Establishing the Program
    - Legal Goals
    - Management Powers and Functions
    - Directly Stated Monitoring Requirements
    - Information Needs of Management Operations
    - Water Quality Criteria
    - Water Quality Standards
    - Permits
    - Compliance
    - Enforcement
    - Construction Loans
    - Planning
    - Water Quality Assessment
3. Ability of Monitoring Systems to Produce Water Quality Information
    - Narrative Information
    - Numerical Information-Data
    - Graphical Information

- Statistical Information
- Average Conditions
- Changing Conditions
- Extreme Conditions
- Water Quality Indices

4. Suggested Information Expectations for Monitoring System
   - Management Information goal(s)
   - Definition of water quality
   - Monitoring system goal(s)
   - Information product of monitoring system
   - Narrative
   - Numerical
   - Graphical
   - Statistical

5. Resulting Monitoring Network Design Criteria
   - Variable selection
   - Site selection
   - Sampling frequency determination

5.4.2 Water Quality Information Specifications

A good design for almost anything from a car to a piece of software includes detailed specifications for the various components. A water quality information system should be no different. Especially important are specifications for the information to be produced.

A number of factors must be considered at the beginning to develop "specs" for the monitoring system design. Long term fixed station monitoring usually seeks to define water quality in terms of averages, trends, and extreme events over time and space. Usually, graphical and statistical methods, along with indices, are best for conveying such information. Special surveys, on the other hand, seek to define water quality conditions which exist at one point or irregularly spaced points in time. Much more "process" detail is typically provided in special surveys than in routine monitoring. Narrative information, actual data, and process models may be used more in this type of monitoring.

Specifications for narrative information could include sample statements as guidelines. Specifications for data themselves would arise from the network design, which would include sampling locations, frequencies, variables, etc. and from the QA/QC program which would specify data quality requirements. Specifications for graphics are fairly straightforward, indicating what types of graphics are to be used and how they are to be constructed.

For statistical information, specifications can be more quantitative. Two categories of specifications are relevant:

(1) Statements of what statistical parameters are important and how precise they need to be. The classic example is a confidence interval on the sample mean with specifications on width and confidence level, which then translate into a required sample size (see Appendix A). Statements of required precision must be subject to continual revision in light of monitoring budget constraints;

(2) Statements of statistical hypotheses to be tested with desired confidence levels and power under specified alternatives (Type I and Type II errors).

An example of such a statement is given by Smith and McBride (1989), who discuss the design of a national water quality monitoring network for New Zealand. They state the following:

> "the intention is to be able to detect a monotonic trend equal to the standard deviation of the detrended data over a five year period, if such a trend occurs."

The implicit null hypothesis is, therefore, that no trend exists for a given variable and location. For design purposes, detectable changes or "detection thresholds" defined as a change in the mean divided by the standard deviation, were calculated for various combinations of sampling frequency, Type I and Type II error.

There will be more discussion of the statistical specifications needed for water quality monitoring in Chapter 6.

5.4.3 Reporting Formats

Information expectations of information users are often tied closely to the form or format in which the information is presented (i.e., the reporting procedures). Thus, it may be highly desirable to prepare example reports and include these examples with the information expectation report. Having information users comment on the reporting formats is one of the best ways to obtain a consensus as to what information the monitoring program is supposed to generate and the form in which it will be presented.

In addition to what the reports will look like, it is most important to specify who will get the reports, how often, and how soon after the data are collected. It is wise to continually ask the question, "Who will use each piece of reported information and how?" If this question can not be answered satisfactorily for a given piece of information, then that information is probably not necessary. This exercise can result in substantial savings and/or redirection of resources to monitoring and information management activities which will provide the greatest value.

This philosophy was utilized in the industrial wastewater monitoring system design described earlier (from Loftis, Porter and Settembre, 1987). Corresponding to each of the seven management objectives listed earlier, a separate monitoring plan was developed (specifying the sampling locations, variables, and frequencies required to meet each information need), and the seven were combined into an overall plan. In the new overall plan, each monitoring activity resulted directly from one of the seven information needs. Those activities which were not derived from a stated objective were deemed unnecessary. As a result the level of monitoring was reduced by at least 20%, and a substantial savings was realized.

The value of monitoring in this case study system was greatly enhanced by automating the generation of routine reports (monthly, quarterly, annual, or date-to-date) using a customized data management system for a PC. While the network design study included the specification of time series analysis and control chart procedures, the reports which proved to be most useful were simple summaries of the form shown in Table 5.1 below.

Table 5.1. Routine data summary for industrial wastewater monitoring.

SUMMARY STATISTICS FOR (START DATE) TO (END DATE) AT (LOCATION) AND (CONTINUOUS, GRAB, COMPOSITE) SAMPLING

	VARIABLE 1 (MG/L)	VARIABLE #2 (MG/L)	VARIABLE #3 (etc.) (MG/L)
No. Observations			
Mean			
Minimum			
Maximum			
Std. Deviation			
No. Excursions			
No. Nondetects			

5.5 DEVELOPING A CONSENSUS ON INFORMATION EXPECTATIONS

The contents of an information expectations report may be controversial, and if the monitoring system designer is to obtain a consensus on the final information product of the monitoring system, the report and the resulting discussions must be handled carefully.

The monitoring system designer must not be strongly tied to the conclusions of the report. If the majority of the information users see the situation differently from the designer, the monitoring goals and objectives should be altered to meet the users' needs and not those of the designer. This point was alluded to in Chapter 1, where the National Academy of Sciences (1977) report was quoted regarding past problems with monitoring system design. The report indicated that designers, not the information users, were determining the information to be produced.

Once an information expectations report is available, it can be presented to information users as a group, explaining its purpose within the total system design.

Following the initial presentation, the designer should visit the various information users (or groups of users) on an individual basis to get their reactions. Based on user feedback, revisions will be required ranging from minor changes to a complete rewrite of the report. It is likely, as the Loftis, Porter, and Settembre (1987) example noted, that it may not be possible to gain a complete consensus on the report. In this case the designer is left to develop the best compromise possible.

The time and effort devoted to preparing an information expectations report and developing a consensus on information goals and objectives will be significant. Time and effort are a part of the total water quality information system design effort and must be budgeted into the initial design planning. Our earlier ballpark estimate for a reasonable design cost for a monitoring system was 10% of the lifetime operating cost. This figure includes the expectations report and consensus building exercise. Designing a water quality monitoring system well-focused on concrete information goals and objectives requires time and money.

6

DATA ANALYSIS

The quantification of information goals and objectives (information expectations), discussed in the previous chapter, makes use of statistics in a number of different ways. In this chapter the authors first review alternative statistical methods available to analyze water quality data, then describe a "protocol" that clearly defines the use of statistics for data analysis and information generation purposes.

Obtaining useful information from data requires some degree of data manipulation and analysis. Specifying the methods or tools for data analysis at the design stage increases the probability that the information generated will be the information needed (and expected). In particular, the information system designer should try to ensure that appropriate statistical methods are selected and that data quality and quantity will be such that statistical results of desired precision can be obtained.

Perhaps the worst mistake that can be made, as noted earlier, is to collect data that may be necessary and then to call in a statistician after the fact to simply "analyze the data and tell us what we need to know." Instead one should call in the statistician first, and ask for help in designing the sampling program from the ground up.

6.1 MATCHING STATISTICAL METHODS TO DATA RECORD ATTRIBUTES

As Ward and Loftis (1986 and 1989) and others have emphasized, statistical methods should match both information needs and attributes of the data records to which they are applied. All statistical methods which produce probability statements, such as confidence intervals, are based on assumptions about the data used in the analysis. Common assumptions include distribution shape, correlation structure, importance of seasonality, or applicability of a linear model.

If the assumptions are not met, the resulting probability statements will be imprecise. In practice, as noted in Chapter 3, it is unlikely or impossible that all assumptions will be met exactly. With any finite sample size from an infinite population, one can never be absolutely certain of the validity of the assumption or of the final conclusions.

However, it is necessary to avoid serious (a subjective judgment) flaws in statistical inferences by avoiding large violations of the assumptions or by using methods which are relatively insensitive to likely violations. Methods which perform well when assumptions are violated are often called "robust."

For water quality applications, graphical and nonparametric (not assuming a normal distribution) methods have proven to be more widely suited to routine data analysis than classical parametric approaches. Many data analysts, consequently, are now recommending that information systems always include these two classes of methods.

However, the authors caution that nonparametric methods do have underlying assumptions with which users should be familiar. These assumptions are often violated in water quality applications, and the implications are important. They also suggest that there are important situations where parametric approaches will work best and that water quality data analysts should be familiar with both classes of methods.

It probably goes without saying that statistical methods should be appropriate for the level of knowledge and experience of the data analyst and that the results produced should match the experience level of the information user. With modern statistical computing capabilities, however, there is a strong temptation to try out lots of data analysis methods without really understanding them. While understanding all the theory is not important, it is essential to understand how the results are to be interpreted and utilized. It is also crucial to understand the most important assumptions of a given method, whether the assumptions are reasonable, and the consequences of violations of those assumptions.

Unfortunately, the precision of statistical information, such as an absolute value of trend or differences in water quality, depends on the variability of data, which is unknown until after the data are collected. Thus it is not known how many samples are needed to obtain a desired level of precision until the monitoring program is already in operation. Initially, therefore, system designers must use historical data, from hydrologically similar locations if none are available locally, to establish initial sampling frequencies. As was pointed out in Chapter 3 (and in Appendix A), a designer can focus on relative trends or differences (expressed in ratio to the unknown standard deviation), without having historical data.

Naturally, as data are collected, sampling programs can be refined and improved. However, there are advantages to maintaining sampling frequencies over the long term for at least some locations. Long term pictures of water quality are greatly clouded when sampling locations are moved. The effect of changing sampling frequency may be less severe. At any rate, one must carefully consider the information tradeoffs involved in continually ''optimizing'' sampling network designs and creating data record limitations in the process.

6.2 ATTRIBUTES OF WATER QUALITY DATA RECORDS

Using water quality data to evaluate water quality conditions in the environment is complicated by a number of data record attributes (Harcum, 1990). Such attributes can be divided into two groups: (1) data limitations; and (2) statistical characteristics. Both groups have an impact on the choice of statistical methods employed.

Data limitations are, for the most part, the result of problems or limitations of the monitoring operations. Common data limitations include:

1. Missing values
2. Sampling frequencies that change over the period of record
3. Multiple observations within one sampling period
4. Uncertainty in the measurement procedures
5. Censoring the measurement signals
6. Small sample size
7. Outliers

Statistical characteristics of a data record result from both the natural variation of water quality in the environment (Harcum, 1990) and from the sampling program itself. Four common statistical characteristics are distribution shape, homogeneity of the variance, seasonality, and serial correlation. While these characteristics do not cause computational problems, they may, as noted above, violate assumptions underlying statistical methods chosen to analyze the data. This, in turn, can result in false conclusions about the behavior of water quality in the environment.

6.2.1 Data Record Limitations

The operation of a water quality monitoring system involves a large number of activities carried out by a number of people. The opportunity for problems is larger than many monitoring system managers want to admit. It is often not until the data are being analyzed that the magnitude of data record limitations is fully appreciated. Recognizing this fact during the design of a monitoring system can help to minimize data limitations during the operation of the monitoring system and to account for these limitations in the data analysis portion of the monitoring system.

A missing value is a measurement period for which no data are recorded. Missing values can occur in a number of ways including equipment breakdowns, lost samples, contaminated samples, poor weather, and employee illness. Missing values can be random or systematic. Systematic missing values can occur when lakes freeze over, wells or streams dry up, or annual reports must be prepared (taking considerable employee time from routine monitoring activities).

Missing values impact both statistical and graphical data analysis methods. Graphical presentations of time series have gaps in the plot of measurements over time. Techniques which compare data over time or space have fewer comparisons available. Statistical tests which require regularly observed sequences are not readily applicable (e.g., standard time series analysis). In general, statistical summaries of the data will be less accurate since all time periods and/or sites are not equally represented (Harcum, 1990).

Changing sampling frequencies often occur when budget restrictions force a reduction in monies devoted to sampling. Shifting water quality problems or a new crisis can cause monitoring resources to be reduced in one monitoring program and shifted to a new program. Statistical techniques which require equally spaced sampling are not directly applicable. Also, statistical summaries over time will be weighted more heavily to the period with higher sampling frequencies unless some data are excluded from the analysis.

Multiple observations in one sampling period cause similar problems. A common reason for this to occur in a water quality data record is when quality assurance/quality control results are "stored" in the same computer record as the original water quality observation. When this occurs, it is an indication that data analysis, as a critical step in any water quality monitoring system, has not been incorporated into the operation of the system. It is often very difficult to go back into a data record and sort out original observations from quality control results. Multiple observations again cause unequal weighting in summaries over time and space. It is often desired to use the "extra" information obtained from the multiple observations, but it is not straightforward as to exactly how this is to be done (van Belle and Hughes, 1984).

Water quality measurements are made with techniques that contain uncertainty. This uncertainty is the result of random analytical error and can vary with the magnitude of the signal and calibration of the measurement equipment. Porter (1986) notes that it is not uncommon to simply ignore measurement error altogether. Measurement error can be greatly magnified by poor quality assurance/quality control (Montgomery, 1987).

Some chemical concentrations in water quality occur close to the analytical detection limit. Trace metals and pesticides are common examples. When the signal from an analytical device is less than a predetermined limit, the data value is typically censored and reported as less than the detection limit (Harcum, 1990). Porter (1986) advocates that the concept of a detection limit be eliminated from water quality monitoring and that the actual signal obtained be recorded along with an estimate of the error. Porter and Ward (1989) describe methods by which the error can be incorporated into the statistical analysis of data reported as a measurement and its associated error.

Censoring may be complicated even further if the detection limit has changed over the period of record. Multiple censoring levels generally occur when different analytical techniques have been employed over the period of record. The changes in techniques are often the result of technological advances being made in the analytical equipment. Multiple censoring also occurs when different laboratory protocols are used or data from different laboratories are analyzed as one data set.

Water quality data records can also contain greater-than (>) data. This occurs with species enumerations and the BOD test. In this case one cannot replace the greater-than value with a number. Statistics based on ranks are the only solution in this case. It should be pointed out that it is not uncommon to find both < and > observations in the same coliform data record.

Small sample sizes may not be a problem per se, but there are limits below which sample sizes cannot be meaningfully analyzed. The choice of techniques may also depend on sample size. As noted in Chapter 3, the power of a test increases with sample size and thus changes over time as data are added to the record.

Both erroneous measurements and extreme events are causes of outliers in a data record. The former is the result of poor quality assurance/quality control while the latter represent correct measurements. It is often difficult, if not impossible, to determine the difference between the two.

These data record limitations should be minimized with a well designed and operated water quality monitoring system and should be addressed in the analysis of water quality data. Recognizing that they can exist and selecting data analysis methods that are robust against them is one way to ensure that the data limitations do not greatly reduce the value of the information being produced by the monitoring system.

6.2.2 Statistical Characteristics of Water Quality Data

For both general understanding and for selecting appropriate methods of routine data analysis, the following statistical characteristics are most important: location (central tendency), variability (scale or spread), distribution shape, seasonality, and serial correlation. The authors discuss each characteristic in turn. A summary of the more applicable statistical methods is presented in Table 6.1.

Location (central tendency) is most commonly determined by the sample mean for normal distributions, sample geometric mean for lognormal distributions (a biased estimator of the true mean, median or geometric

Table 6.1. Methods for characterizing data.

DATA CHARACTERISTIC	METHODS	REFERENCE
Central tendency	Sample mean	Gilbert (1987)
	Sample median	"
	Sample geometric mean	"
	Box-and-whiskers plot	Ward et al. (1988)
Spread	Sample standard deviation	Snedecor and Cochran (1980)
	Range, maximum-minimum	Conover (1980)
	Interquartile range	"
	Box-and-whiskers plot	Ward et al. (1988)
Distribution shape	Histogram	Snedecor and Cochran (1980)
	Sample Skewness	"
	Sample Kurtosis	"
	Chi-Squared test	"
	Kolmogorov-Smirnov test	Zar (1984:92)
	Shapiro-Wilk test	Gilbert (1987)
Seasonal variation	Time series plots	Ward et al. (1988)
	Seasonal box-and whiskers	"
	ANOVA	Snedecor and Cochran (1980)
	Kruskal-Wallis test	Conover (1980)
Serial correlation	Sample autocorrelations	Box and Jenkins (1976)
	Rank von Neuman test	Gilbert (1987)

mean—Gilbert, 1987:172), or sample median for any distribution. For symmetric distributions the mean and median are equal. Spread is most commonly described by the sample variance or standard deviation.

The standard deviation, as a measure of spread, is particularly appropriate and meaningful for the normal distribution and may be estimated for any data set, regardless of distribution. However, in the general case, the interquartile range or middle 50% of the data may be a more meaningful sample statistic. The median and interquartile range are frequently used as summary statistics along with nonparametric tests when it is uncertain as to whether the data are normally distributed—the usual case in water quality applications. The box-and-whiskers plot mentioned in Chapter 5 presents several measures of location and spread at once—sample median, interquartile range, maximum and minimum, and/or extreme percentiles such as the 5th and 95th.

A relative frequency plot or histogram is probably the most straightforward presentation of distribution shape and should be included as an early step in data characterization. Normal (or other distribution) probability plots are often used along with histograms to help determine whether a given data set is drawn from the specified distribution. Distribution shape may also be described by the sample skewness coefficient and sample kurtosis (Snedecor and Cochran, 1980). Skewness is a measure of the asymmetry of a distribution. Symmetric distributions, like the normal, have zero skewness. Distributions skewed to the right, such as the lognormal distribution, have positive skewness. Distributions skewed to the left have negative skewness. Kurtosis describes the "peakness" or "flatness" of a distribution. The normal distribution has a kurtosis of 3.0. Many formal tests of normality are available. The more popular have, in the past, included the Chi-Squared Goodness-of-Fit and Kolmogorov-Smirnov tests (Conover, 1980 and Sanders, et al. 1983). Newer, more powerful tests were mentioned in Chapter 3. The Shapiro and Wilk W-test (Gilbert, 1987) is a good choice as long as the sample size is less than 50. A simple and powerful alternative is to test for significance of the sample skewness and kurtosis (Snedecor and Cochran, 1980). If the sample skewness differs by too much from zero or the sample kurtosis differs by too much from 3.0, then the data would be regarded as non-normal. Since many parametric statistical methods work for the general class of unimodal (single-peaked) symmetric distributions, an inspection of the histogram followed by a test of the skewness coefficient is adequate for many applications.

It is well to remember at this point that acceptance of a null hypothesis of normality does not necessarily mean that the data are normal. In fact, it is almost inconceivable that a real data set would be *exactly* normally distributed. Accepting the null hypothesis indicates that there are not enough data to reject it.

Thus in this, as in most, hypothesis testing applications, one is faced with a subjective decision of how many data points are enough before one can have confidence in the validity of a conclusion. In choosing between parametric vs. nonparametric methods, one is also faced with the question of how normal the data must be to justify parametric methods. For the two-sample t-test, this question was examined in detail by Montgomery and Loftis (1987).

As noted earlier, though, for routine data analysis, one should look to nonparametric methods in general and use parametric methods only in special situations. Such situations would include regulatory requirements to use parametric methods, acceptance of a null hypothesis of normality based on a large background data set, or lack of availability of a powerful nonparametric method. In the latter case, one should consider data transformations, such as the log transform, if necessary to achieve near-normality before applying the parametric test. Before taking this course of action, however, one should read Gilbert's (1987:149) well-stated misgivings about use of transformations. One should also investigate the consequences, such as low power or high false rejection rate, of applying the parametric test to non-normal data.

Seasonality is a predictable change in location and/or spread from one time of year to another and is the general rule, rather than the exception, in water quality—especially surface water and shallow alluvial ground water. Seasonality is important to consider in water quality data analysis since it violates the assumption of independence in time inherent in many statistical prodecures. Furthermore, it would result in an inflated estimate of variance when observations from more than one season were included.

Seasonality may be observed visually in time series plots or seasonal box-and-whisker plots as described in Chapter 5. A formal test for seasonality may be performed by testing a null hypothesis that the means or medians of all seasons are equal against the alternative that they are not all equal. For normal distributions, analysis of variance (ANOVA) is appropriate (Snedecor and Cochran, 1980). However, the equivalent nonparametric Kruskal—Wallis test (Conover, 1980) is more appropriate for general use in water quality. Both methods test for equality of location across multiple groups and have broader applications in water quality than simply testing for seasonality.

Autocorrelation of a time series represents a carry-over of information from one observation to the next. Positive autocorrelation is the tendency of high values to follow high values or low values to follow low values. Negative autocorrelation is the tendency of high values to follow low values and vice versa. Autocorrelation in the absence of (or after removal of) seasonality or trend is referred to as serial correlation.

Serial correlation, like seasonality, violates the assumption of independence inherent in many statistical tests. Unfortunately, while tests can usually be easily modified to account for seasonality without loss of power, the same is not true for serial correlation. Since serial correlation and trend are very similar and sometimes indistinguishable, trend tests which correct for serial correlation will be less sensitive to changes (less powerful) than those which do not.

Serial correlation is often characterized by the sample autocorrelation function $r(k)$ of a series with seasonality and trend removed. The sample autocorrelation coefficient $r(k)$ is simply the correlation coefficient between observations k time periods apart. The term k is referred to as the lag. A simple test for serial correlation can be performed by comparing the computed values of $r(k)$ for the first few lags against standard error limits for an uncorrelated series. Box and Jenkins (1976) give $1/\sqrt{N}$ as the approximate standard error for an uncorrelated series where N is the number of data points used to compute $r(k)$.

If working with a normally distributed time series, $r(k)$ is approximately normally distributed with mean zero for uncorrelated observations. In this case, a 95% two sided test of the null hypothesis that the series is not serially correlated amounts to comparing $r(k)$ against the limits $\pm 1.96/\sqrt{N}$ for all k. In water quality application, lag one is usually most important. It is also important to remove seasonality and trend before performing the test unless one wants to reject the null hypothesis if the data are serially correlated *or* seasonal *or* have a trend.

More generally applicable tests for serial correlation, which do not rely on normality assumptions, are based on the ranks of data rather than the actual observations. For example, one could use a test of Spearman's rho, instead of $r(k)$ (Conover, 1980) or the rank von Neumann test described by Gilbert (1987).

6.3 METHODS FOR ROUTINE DATA ANALYSIS

As suggested earlier, routine data analysis is most often directed toward obtaining information on average conditions, changing conditions or trends, and extreme conditions. The statistical methods which are used to obtain this information may be classified as graphical procedures, estimation procedures, or tests of hypotheses. The authors shall mention examples of each type, for each of the three general information objectives (see Table 6.2). Some of the relevant techniques are also applicable to characterizing data and have been discussed in preceding sections.

6.3.1 Average Conditions

As mentioned earlier, average conditions are most commonly defined by the sample mean, sample geometric mean or sample median. It is generally desirable to obtain these estimates as intervals with an associated confidence level as opposed to a single point estimate. The calculation and use of confidence intervals for the sample mean and/or geometric mean are discussed in most statistical texts such as Snedecor and Cochran (1980) and in Sanders et al. (1983) for water quality in particular. Two complicating issues for computing means are seasonal variation and serial correlation. The former is handled in a straightforward manner by computing means (or medians etc.) for individual seasons as opposed to overall or annual means. The latter issue is more complicated.

Serial correlation tends to reduce the effective independent sample size. Thus the width of the confidence interval on long-term means for a given number of samples is increased, compared to that for independent observations. Both Sanders et al. (1983) and Gilbert (1987:38) discuss the issue in considerable detail. Gilbert (1987:164) also contains a discussion of characterizing lognormal populations in terms of the mean, geometric mean, median, quantile points and variance. Sachs (1984) presents many expressions for confidence intervals including those mentioned above and expressions for confidence intervals for the variance and coefficient of variation of normal populations.

6.3.2 Trends or Changing Conditions

When evaluating changes in water quality over time (or space) one is usually interested in one of two types of change. The first type is differences in location and/or in spread between two or more groups of data. The groups could represent either different times or different locations. The second type of change is a gradual shift in level (mean, geometric mean, or median) occurring at a given location over time; for example, an approximately linear trend.

Table 6.2. Methods for routine data analysis.

INFO TYPE \ METHOD TYPE	GRAPHICAL	ESTIMATION	TESTING
Average Conditions	Box-and-Whisker Plots Time Series Plots	Sample Mean, Median or Geometric Mean with Conf. Limits (Gilbert, 1987) Sample Std. Dev. with Conf. Limits (Sachs, 1984) Seasonal Hodges-Lehman Estimator (Hirsch, 1988)	Student's t Test Paired t Test ANOVA (Snedecor and Cochran, 1980) Mann-Whitney Test Wilcoxon Signed Rank Test Kruskal-Wallis Test (Conover, 1980)
Trends or Changing Conditions	Annual Box-and-Whisker Plots Time Series Plots	Linear Regression (Snedecor and Cochran, 1980) Seasonal Kendall Slope Estimator (Gilbert, 1987)	t-Test for Significance of Slope (Snedecor and Cochran, 1980) Seasonal Kendall Test (Gilbert, 1987)
Extreme Conditions	Time Series Plot with Excursion Limit	Proportion of Excursions (Ward et al. 1988)	Confidence Limits on Proportions (Gilbert, 1987) Test for Equality of Proportions (Snedecor and Cochran, 1980) Tolerance Intervals (Conover, 1980)

Comparing Two or More Groups of Data

The first type of change is evaluated by estimating the location and/or scale parameters of each group of interest, perhaps using interval estimates. In addition, one may wish to perform formal tests of hypotheses regarding the significance of apparent difference between the groups. Many tests for comparisons between groups are suggested in statistical texts.

The classic test for equality of means between two groups of normally distributed observations is the Student's *t*-test (Snedecor and Cochran, 1980). When observations may be paired, as in the optimal impact study mentioned in Chapter 3, a paired t-test will provide higher power.

The nonparametric analogs of the *t*-tests are the Mann-Whitney test for independent observations and the Wilcoxon signed rank test for paired observations (Conover, 1980). For estimating the difference in medians between two groups of seasonal data, the seasonal Hodges-Lehman estimator appears to be the best alternative (Hirsch, 1988).

When more than two groups of data are involved, analysis of variance is appropriate for testing for equality of means in normally distributed data (Snedecor and Cochran, 1980). The nonparametric analog of ANOVA is the Kruskal-Wallis test, (Conover, 1980 or Gilbert, 1987:250). The Krukal-Wallis test or ANOVA has been mentioned previously as appropriate for testing for equality of medians (or means) across seasons. These tests are also appropriate for comparing data from multiple wells, for example comparing a single compliance or detection well to multiple background wells (U.S. Environmental Protection Agency, 1989). Such a comparison would be most valid when data from the background wells and detection or compliance well were collected for the same time period (as in an optimal impact study).

Testing for Trend at a Given Location

The authors use the term trend to refer to a fairly gradual, monotonic, increase or decrease in the mean of a water quality variable over time. The classical method of trend analysis is estimation of a linear trend slope against time using least squares regression, followed by a *t*-test of the significance of the slope parameter (Snedecor and Cochran, 1980). When using this approach it is usually desirable to deseasonalize the data, by removing seasonal means from each observation, before performing the test.

A much more widely applicable test is the nonparametric seasonal Kendall test originally described by Hirsch et al. (1982) and also found in Gilbert (1987:225). The seasonal Kendall test is an extension of the Mann-Kendall test for trend (also described in Gilbert, 1987:208). The Mann-Kendall test may be applied to data which do not vary seasonally, and the seasonal Kendall test handles seasonality in the mean very effectively. The seasonal Kendall test also performs reasonably well when the both the mean and variance are seasonal (Loftis and Taylor, 1989). For estimating the magnitude of trends, the seasonal Kendall slope estimator may be used (Gilbert, 1987).

Positive serial correlation tends to cause tests for trend to reject the null hypothesis more often than they should when in fact, there is no trend. That is, the actual significance level of the test will be greater than the nominal or assumed significance level. Corrections for trend tests have been proposed, and the most widely applicable is probably the corrected seasonal Kendall test described by Hirsch and Slack (1984). Unfortunately, correcting for serial correlation greatly reduces the power of the test, especially for short data records of ten years or less (Hirsch and Slack, 1984; Loftis and Taylor, 1989; Loftis et al. 1989a).

Since changes in means over time may be of concern whether they result from serial correlation or deterministic trend, the authors generally recommend using the uncorrected seasonal Kendall test as a screening tool. When the test results in a rejection of the null hypothesis of no trend, one can then look further into the probable causes. If necessary the serial correlation structure of the series may be investigated as a possible explanation for an apparent shift in level.

6.3.3 Extreme Values

Extreme values or excursions are most commonly defined as observations which exceed any preset limit, such as a stream standard or drinking water standard, compliance limit, action limit, or control limit. Excursions may be evaluated simply by comparing each value against the standard. As discussed in Chapter 3, standards compliance is sometimes measured in terms of the fraction of time or fraction of observations in which a standard is not met. This fraction or proportion is estimated simply by dividing the number of excursions by the total number of observations. The proportion of excursions may also be observed visually by marking the excursion limit as a horizontal line on a times series plot of the data and noting the fraction of time that the limit is exceeded. Gilbert (1987:142) and many other texts provide equations, graphs, or tables for calcuating confidence limits for proportions. Chapter 3 discussed tests of hypothesis on the compliance propórtion.

One might also wish to test for equality of proportions between two groups of data. For example one might formulate a null hypothesis that the proportion of detectable values in a compliance well is equivalent to the proportion of detectable values in a background well or group of background wells. Such tests are suggested by the U.S. Environmental Protection Agency (1989) for data sets with large numbers of nondetects. Snedecor and Cochran (1980:121) describe tests for comparing proportions with both paired and independent samples. Methods for estimating sample size requirements are also discussed.

In some cases, excursion limits are defined in such a way that there is a small probability of an observed value falling outside the limits unless a change has occurred in the variable of interest. Such limits are often called control limits and are established using a historical data record.

A type of control limit of particular interest is the tolerance limit. Tolerance limits, widely used in quality control, define an interval which contains a certain proportion of the distribution, p, with specified probability q. Thus if one establishes a tolerance interval with $p = q = 95\%$, there is at least a 95% probability that 95% of future observations will fall within the interval—if no change occurs in the random variable we are tracking. If more than 5% of future observations fall outside the interval, one would infer that a change had likely occurred.

Statistical tolerance limits may be calculated using a normality assumption (Bowker and Lieberman, 1972; Sachs, 1984); or nonparametric intervals may be found (Conover, 1980; Sachs, 1984). Unfortunately, the nonparametric intervals are impractically wide for many applications, especially those with small sample sizes. This is one of the few cases where a parametric method might be considered over nonparametric methods for routine application in water quality. To handle skewed-right data, one should consider a log transform to achieve a semblance of normality before computing a tolerance interval. The logs of future observations would then be compared against the tolerance interval computed from the logs of background data.

Loftis, Harris and Montgomery (1987) discuss the use of tolerance limits for detecting changes in ground water quality at regulated facilities. Tolerance intervals are also recommended for analysis of RCRA monitoring data by EPA (1989).

6.4 DATA ANALYSIS PROTOCOLS

The translation of water quality monitoring objectives into meaningful guidance for data analysis can best be accomplished within the context of a protocol. While the concept of a "protocol" has existed in sampling and laboratory analysis for many years, it is only recently being considered for the data analysis end of the water quality monitoring system. The creation of a data analysis protocol will force a careful translation of information objectives into data analysis methods which can be audited to insure that the required information can be obtained. The term "data analysis protocol," as it is used herein, implies that the procedures for data analysis, associated with the ongoing operation of a water quality monitoring system, will be described in a step-by-step manner along with the reasons for each step. The description of the procedures will begin with preparation of the data for analysis and end with an interpretation of the water quality picture which can be drawn from the monitoring system's results.

It is one thing to say that a monitoring system is needed to provide information on trends in water quality, for example, and another to spell out exactly how the data are to be analyzed to yield such information. That is the purpose of a data analysis protocol—to describe the graphical, statistical, or other means (such as a water quality index) to be used in obtaining the desired information from a water quality monitoring system. The protocol ties together the last three portions of the monitoring system as defined in Chapter 2: data analysis, reporting, and information utilization.

A data analysis protocol (DAP), carefully and thoroughly relates information objectives to a specific graphical and statistical means of obtaining the information. Unfortunately, such an effort is often not undertaken until after a monitoring system has been in operation for a number of years and the data are in need of analysis. However, including a documented data analysis protocol within the system design can greatly enhance the performance of a total water quality monitoring system. It ensures consistency in analysis when data may come from different origins and are analyzed by different people. A well documented DAP also can support a higher level of data analysis by personnel who do not necessarily have advanced level training in the field of statistics (Martin et al. 1988).

If questions arise as to whether the chosen methods for data analysis are appropriate or correct, a DAP permits a complete audit of the entire procedure by qualified professionals. There is an advantage to having arguments about data analysis methods aired before data are collected and analyzed. If no DAP exists, arguments will almost inevitably develop regarding the information results and data analysis methods. In such situations it is very easy to find fault with just about any statistical procedure that has been applied, and it is also difficult to defend a particular data analysis approach chosen after the data are collected. This is not to say that a DAP will eliminate the possibility of future arguments over the information results of a monitoring system, but as a part of an initial design it becomes a defensible approach since it was initially agreed upon by both monitoring system personnel and the users of the resulting information.

A data analysis protocol has an advantage in that it can incorporate existing knowledge about the general behavior of water quality variables. This may permit the use of assumptions about such behavior before a data record is long enough to confirm those assumptions. In turn this permits, with care, the analysis of shorter records to obtain the desired information.

The existence of a DAP permits "quality assurance" activities to follow the flow of information into the data analysis and reporting phases of the monitoring system. The person doing data analysis can now cooperate with the quality assurance staff to demonstrate that the information goal is in fact being met. Without a DAP, quality assurance efforts often end at the exit of the laboratory or perhaps at the database management point in the monitoring system—rarely do they follow the data into the analysis phase of the monitoring system.

There are, of course, disadvantages to a DAP. A DAP can become too routine and inhibit true exploration of a data set for a new understanding of the behavior of water quality in the environment. However, a DAP can include both exploratory and confirmatory techniques (McLeod et al. 1983). Also, periodic evaluations and revisions of a DAP can be conducted in light of the evolving state-of-the-art in water quality data analysis.

The role of a data analysis protocol is crucial. Its absence creates a barrier to the flow of information shown in Figure 2.1. Its presence connects raw data directly to the information goals for which the data are collected in the first place.

6.5 CONTENTS OF A DATA ANALYSIS PROTOCOL

A data analysis protocol must, at a minimum, contain five components:

1. A statement of the exact information to be produced;
2. Procedures to be used to prepare a raw data record for graphical and statistical analysis;
3. Means to visually summarize the behavior of the water quality variables;
4. Recommended statistical methods which yield the desired information; and
5. Reporting formats for the resulting information.

In preparing a DAP, it is the monitoring system designer's goal to spell out each component in sufficient detail that there is little chance for confusion. This is the same goal associated with protocols for collecting samples and analyzing them in the laboratory.

6.5.1 Statement of Information to be Produced

The DAP should be based on information expectations already developed as part of Step 1 in the design process recommended in Chapters 4 and 5. As noted earlier, there are three basic types of graphical and statistical results that one can expect to obtain from the analysis of water quality data:

1. Estimates of *average conditions;*
2. Indications of *changing conditions;* and
3. Descriptions of *extreme conditions.*

Specific information in the above three categories must be related to the general questions asked of a water quality monitoring system such as those noted at the beginning of Chapter 3. To accomplish this Ward (1989) and Ward and Loftis (1989) suggest preparing the following:

1. A statement of management's goal with respect to water quality conditions

2. A statement of the information on the behavior of water quality variables that best relates to the management goal. Does information on average, changing, and extreme conditions best relate to measuring achievement of management's goal?
3. A clear definition of water quality which includes the variables and "population" to be measured
4. A description of the statistical methodology to be used to analyze the data to obtain the information sought in 2.
5. A list of the statistical parameters to be estimated or statistical hypotheses to be tested along with the confidence and power levels sought
6. Interpretation of the data analysis results relative to their water quality implications
7. Interpretation of the water quality behavior relative to management goals.

Ward and Loftis (1989) illustrated this format, in an extremely brief fashion, for the specific management goal of maintaining or improving water quality, as follows.

Management Goal:	Maintain or improve water quality
Monitoring Goal:	Detect trends in water quality
Definition of Water Quality:	Variables listed in standards
Statistical Methodology:	Kendall Tau rank correlation test at 90% with Sen Slope Estimate
Statistical Hypothesis:	Null hypothesis: No trend exists Alternate hypothesis: Trend exists
Data Analysis Result:	Conclusions regarding existence of trends in the data with estimates of trend magnitude
Information Product:	Management goal is met when all slopes are zero or indicate improvement

The case studies in Chapter 9 provide additional examples.

6.5.2 Data Preparation for Analysis

As noted earlier several types of data record limitations affect statistical analysis. These include:

1. Missing values
2. Different sampling frequencies
3. Multiple observations
4. Measurement uncertainty
5. Censored data
6. Small sample size
7. Outliers

Differences in how these limitations are handled can result in differences in information produced. A major component, therefore, of a DAP is specifying exactly how these data limitations are to be addressed before data analysis proceeds. One way to handle this issue is to modify a "raw" data file to create a "data analysis" file that is ready for graphical and statistical analysis.

To illustrate consider the DAP described by Ward et al. (1988) for a groundwater quality monitoring system. In preparing data for analysis, the DAP first specifies the following:

1. Missing values are left as blanks in the record—no effort is made to insert an estimated value.
2. A set sampling frequency of one observation per quarter is chosen. (Besides being a typical sampling frequency for groundwater monitoring systems, the problems of serial correlation are reduced at this frequency.) If multiple observations occur within each time interval (quarter), they are averaged to obtain a single number.
3. Censored data (<) are entered as one-half of the detection limit.
4. No statistical techniques are used to remove outliers. Instead, Ward et al. (1988) recommend that each questionable observation be carefully investigated. If a documented reason for removing it is discovered, the outlier is to be removed. Otherwise, the number remains in the record.

Using these four "rules," a raw data record is converted to a file ready for data analysis.

6.5.3 Graphical Analysis

In general, it is recommended that a first step in data analysis would be to look at the data in a graphical format in order to gain an understanding of water quality behavior which can be used in interpreting statistical results. A graphical presentation of the data can be used to help ensure that statistical methodologies are being used correctly. For example, homogeneous variance (contant variance over time) is assumed for many statistical calculations. If a plot of the data reveals that this is not a valid assumption, it may be necessary to break the record into portions for which the assumption holds. If a graphical presentation reveals that the data contain a highly seasonal pattern, this information may be helpful in deciding whether a particular statistical method needs to be modified before application, or another method needs to be selected.

There are a number of procedures for graphing water quality data. Ward et al. (1988) recommend, as part of their groundwater data analysis protocol, that the data first be viewed in a time series plot followed by seasonal and annual box-and-whisker plots. Examples of these were presented in Figures 5.1 and 5.2.

Leo et al. (1984) and Rickert et al. (1975) are two examples of reports that describe water quality along a river using plots of concentrations versus river miles. Smith et al. (1987) and Smith et al. (1982) graphically describe water quality trends using a map with arrows pointing up for increasing concentrations and arrows pointing down for decreasing concentrations. Circles were used to indicate no change. Everett and Schmidt (1979) and Ward et al. (1989), proceedings of symposia on water quality monitoring, contain additional examples of graphical displays of water quality data.

At this time, no one method of graphical display appears to be universally accepted. Furthermore, these past examples must not be viewed as the final word on the subject. The information revolution taking place in society today is creating many new technologies and procedures for presenting information graphically. For example, the *USA Today* newspaper has developed a graphical approach to presenting data that we in water quality could perhaps better utilize. Likewise, the *Wall Street Journal,* in presenting economic data, presents raw data along with graphics to summarize the information contained therein. Tufte (1983) is an excellent review of methods for visually displaying a wide variety of quantitative information.

As Lettenmaier (1988) notes, the difficulty with most water quality assessments, especially large scale trend assessments, is the interpretation and presentation of the massive volume of results. He reports that for the 35 water quality variables suitable for trend assessment at the approximately 500 NASQAN sites, there are about 15,000 individual trend test results. This situation has led others to consider the development of a water quality index as a means of integrating water quality information in a form that can be easily presented graphically.

6.5.4 Statistical Analysis

In actually estimating water quality statistical parameters or in testing hypotheses, specification of exact methodologies becomes critical to obtaining consistent and correct information about water quality. The area of water quality statistics has been developing over the past 10–15 years, and, while there have been many quantitative developments in the field, there is still considerable debate about the "best" statistical methods to use in various applications.

Thus, in specifying the statistics to be used in a data analysis protocol, there will have to be considerable judgement utilized on the part of the monitoring system designers. The key point, however, is that the designer attempts to define the means of data analysis prior to collecting the data, thereby increasing the integrity of the information system.

As mentioned earlier, the selection of methods should match, as closely as possible, the statistical characteristics of the data being analyzed. We touched upon three characteristics of water quality data which are particularly important: distribution shape (normal vs. non-normal), seasonal patterns, and correlation in time.

In parallel with the general recommendations above, a data analysis protocol can handle the first characteristic quite easily by using graphical and nonparametric methods as opposed to those which assume a normal distribution. The next issue is resolved by assuming that water quality data are usually seasonal and including only those methods which handle seasonality.

By making these decisions when the protocol is developed, the data analyst is freed from the task of analyzing monitoring data as they are collected to determine whether assumptions of normality or seasonality are met. Instead the assumptions can be confirmed. Furthermore, variability in output information will be minimized since all analysts will use the same methods. This is not to say that parametric methods should never be specified in a DAP, only that graphical and nonparametric methods will be much more useful for routine applications.

The last issue, temporal or serial correlation, is much more difficult to handle effectively in routine data analysis. Temporal dependence, other than that caused by seasonality, is redundancy in information from one observation to the next and violates the assumption of independence inherent in most statistical tests. Unfortunately, convenient methods to identify and handle this situation, especially for small data sets, are not available.

The recommended approach for dealing with serial correlation in routine monitoring and data analysis is (1) to generally avoid specifying sampling frequencies which would likely produce appreciable serial correlation and/or (2) to average data over time periods long enough to reduce the level of serial correlation.

Some monitoring objectives, such as detection of excursions or standards violations, may require very frequent or even continuous monitoring. For such purposes, the issue of serial correlation is moot. When the closely spaced data are analysed for another purpose (such as trend detection), however, appropriate treatment (such as averaging prior to trend analysis) will be necessary to deal with serial correlation.

Since no methods for dealing with serial correlation appear to be completely satisfactory for routine DAPs, data analysts should be aware

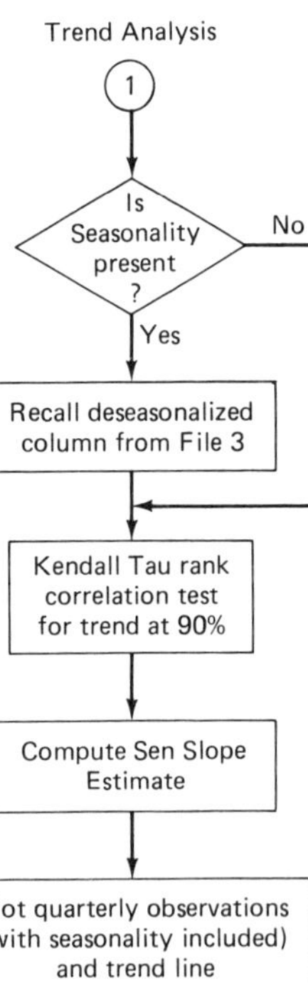

Figure 6.1 Flowchart for trend analysis portion of protocol.

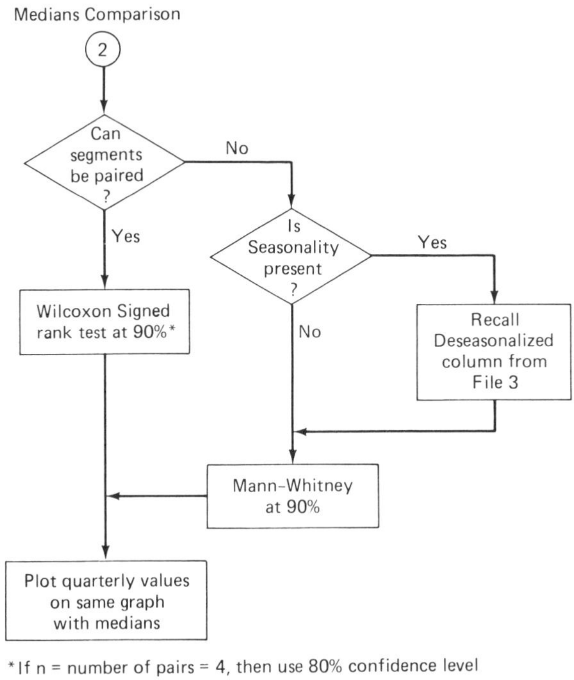

Figure 6.2 Flowchart of medians comparison portion of protocol.

of the possible effects of serial correlation on the validity of statistical conclusions.

Statistical analysis of water quality data involves selecting a parameter to be estimated or a hypothesis to be tested and then choosing the statistical approach. The literature has recorded many opinions on the parameters and tests of importance and also many opinions of which statistical methods should be employed to do the calculations. It is not possible here to state with certainty which methods are best for all water quality monitoring systems. However, Tables 6.1 and 6.2 summarize some of the more broadly applicable statistics for water quality data analysis.

In the groundwater DAP proposed by Ward et al. (1988), statistical methods were selected for testing for trend, medians comparisons, and excursion analysis. Flowcharts describing the sequence of applications are presented in Figures 6.1 through 6.3. Such specifications clearly state how various statistical methods are related to each other and to the overall information goals.

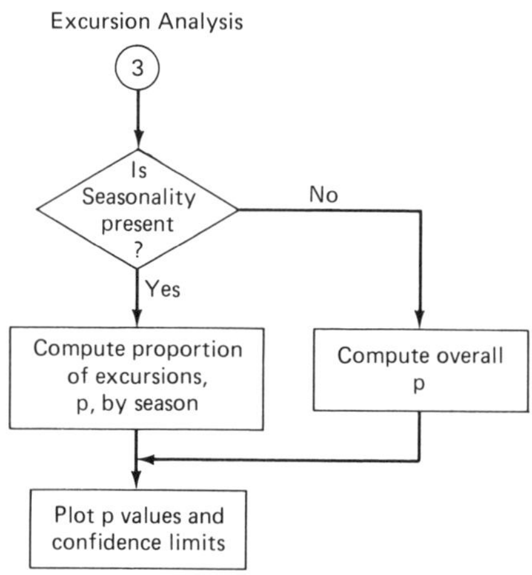

Figure 6.3 Flowchart of excursion analysis portion of protocol.

6.5.5 Reporting

Reporting formats should be specified in the DAP with illustrative examples included. Given the lack of agreement in the water quality monitoring profession as to how reports are to be constructed, the designer of a monitoring system will have to put considerable creative effort into this step.

Ward et al. (1988), in their groundwater DAP, suggested the format presented in Figure 6.4 using appropriate symbols to provide very concise information. Another example, using both verbal and graphical descriptors of water quality within the United States is presented by the Association of State and Interstate Water Pollution Control Administrators (1984).

6.6 SUMMARY

Data analysis protocols are evolving in water quality monitoring as the need for consistent and auditable means of analyzing data is becoming increasingly apparent. There are no set procedures that address all the information needs in water quality management. However, reasonable data analysis protocols can be suggested for use with given types of water quality monitoring. Such protocols, however, must recognize the evolving nature of water quality data analysis and be flexible enough to adapt to future developments in applications of statistics to water quality.

			Medians			Medians				
	Dates		Quarters	Annual	Segment		Comparisons		Trend	Excursions (Proportion of Observations ± Conf. Interval)
Well Number	From	To	1 2 3 4	(Yr)	Total	Well No.	Time Seg.	=, < >	(+, −, 0)	

Figure 6.4 Suggested reporting format for data analysis results.

7
NETWORK DESIGN

At this point in the design process, Steps 1 and 2 from Chapter 4 have been completed, thereby defining the information to be produced and the data analysis methods to be used. Progressing to Step 3 moves the design process into the "field." Questions of interest now include: "Where will samples be collected?," "What will be measured?," and "How often will samples be collected?". Such decisions must be carefully related back to the information expectations or design criteria specified in Steps 1 and 2.

Step 3 is referred to as "network design" because it defines, among other things, a network of locations where the sampling will take place. As noted earlier in the book, network design was deemed, for many years, to be all that was required before sampling could begin. Networks were established, generally, on an ad hoc basis. Locations were often selected based on ease of access in the opinion of a single individual or a few people sitting around the table. Sampling frequencies were often defined by operational convenience. Variables selected for measurement were also decided by people sitting around a table and ability of the laboratories to conduct the measurement. The where, what, and when decisions were then used to direct the operation of the sampling and laboratory operations. The reasons behind the "where, what, and when" of sampling and laboratory operations were rarely documented. Nor were these decisions related, in a quantitative manner, to any predefined information goal for the total monitoring system.

The authors are not suggesting that people should not sit around a table and discuss the elements of water quality monitoring system design. They must, however, specify the design of the entire monitoring system and not just the network. Thus, a key element of network design, is to document the relationships between the where, what, and when of sampling and the information goals of the total system.

During the early development of monitoring system design methodologies, the where, what, and when topics of network design dominated the literature. Reports by Beckers and Chamberlain (1974), Kittrell (1969), Lettenmaier (1975), Montgomery and Hart (1974), Sanders (1974), Sherwani and Moreau (1975), and Ward (1973) address the problem of monitoring network design primarily from the point of view of where to sample, what to measure and how often to measure. Ponce (1980) captured many of the evolving concepts in one document published by the U.S. Forest Service. Sanders et al. (1983), as part of a course taught on the subject of water quality network design, also provided a review and synthesis of available literature while adding new material developed as part of their research and teaching program at Colorado State University.

Many of the procedures described in Sanders et al. (1983) are used today for network design, but the better defined information goals of today permit better documentation of the reasons for the where, what and when of monitoring. In particular, the information expectations report defined in Chapter 5, the "why" of monitoring, becomes an excellent basis for quantification of the where, what, and when of monitoring system design.

7.1 LOCATION OF SAMPLING SITES

Location of a sampling site, whether it is a permanent site in an ambient monitoring system or a temporary site in a special survey, is extremely critical to the ability of the information system to provide information on the target water quality population. If the sample is not representative of the target water quality population, then the rest of the information system is producing information on quality of the wrong water. Thus, not only must the target water quality population be identified in the information phase of the design (Steps 1 and 2), but once identified, considerable science must be brought into the design process to ensure that the target population is in fact sampled in a representative manner.

The target water quality population can vary from the surface water quality of an entire country to the effluent from a particular industry's discharge. During Steps 1 and 2, in preparation of the information expectations report, this population has been clearly defined, documented and agreed upon by the information users. The specific sites for sampling must now be identified.

The number of sampling sites to be employed is usually a function of the budget available for sampling, the size of the water quality population, and the variation in water quality within the target population. Normally, an analysis of the variation in quality over a length of river, area of a lake, or within an aquifer will give an indication of the number of sampling sites needed to characterize a water quality population. To obtain a preliminary understanding of such variation in quality, it may be necessary to perform several investigations, especially if such knowledge has not already been obtained through previous monitoring.

In the case of ground water monitoring, preliminary monitoring is very expensive. Thus developing the final network design may involve installing a few wells, observing variations in both geology and water quality, applying ground water flow models or others as appropriate, and adding additional wells as needed. Uncertainty will be a constant companion in the process of locating ground water quality sampling wells in order to obtain "representative" samples from an aquifer.

Identifying sampling sites requires several levels of specification. First, the macro level defines the river reach, the area of a lake or estuary, or the portion of an aquifer to be sampled. These macro level locations are generally a function of the information objectives. The micro level defines the exact places within the river reach, lake or estuary, or ground water to be sampled. The micro location is determined by the need to obtain a "representative" sample within the macro location.

For example, within an identified river reach for which water quality information is required (the macro location), a representative sample may be determined by the location where full mixing of inflows has taken place. (In the past the term "completely mixed" was often used to describe a point where a representative sample can be taken. This term may be misleading because it is not possible to have a completely mixed cross-section due to the differential advection caused by the greater speed of the water in the center of the channel as compared to that along the bottom and the banks.) At the point where a full mixing of inflows has occurred, it would be possible to obtain a representative sample of the water quality population of that particular reach of river. In some cases, because of the proximity of inflows, such a spot may not exist, and multiple samples in the cross-section may be needed. These can then be bulked to obtain a single sample, or each can be analyzed separately and the results combined.

Procedures for mixing calculations are given in Sanders et al. (1983) and McBride, Rutherford and Pridmore (1988). In a lake, estuary, or ground water quality situation, where there is no one dominant space coordinate, finding a completely mixed point for sampling may be an inappropriate concept. For instance, in monitoring lake phytoplankton one must recognize that their type and abundance can vary considerably through the depth and between sites. It may be necessary to take a number of samples from open-water sites (Pridmore 1987). For most purposes there is no need to analyze each sample independently. The samples can be combined and analyzed.

Sanders et al. (1983) describe a number of ways to determine macro location and the representativeness of a sample. These methods will not be repeated here. The reader is referred to Chapter 4 in Sanders et al. (1983) for details of methods for locating sampling sites within the information goals established in Steps 1 and 2 and as documented in the information expectations report described in Chapter 5.

7.2 SAMPLING FREQUENCY

Specifying the frequency for sampling within a water quality monitoring system requires integrating a number of factors. These include:

- Information sought (such as magnitude of trend to be detected or degree of standard compliance, including the agreed error risks)
- Statistical methods employed to obtain the information
- Statistical characteristics of the water quality "population" being sampled
- Budget available to support travel to sampling sites
- Distance of sampling sites from the laboratory
- Number of sampling sites within the network
- Ability of laboratory to process samples

The latter four involve operational constraints on the monitoring system that do not relate to the information goals. It is these latter four that were, for many years, overriding factors in the design of water quality monitoring networks. During the past 10–15 years, however, the first three factors above have come to the fore and are now viewed as most important, while the latter four are viewed as constraints. Thus, the tradeoffs between the operational constraints, especially budget, and information obtained can be evaluated.

Once the information goals, data analysis methods, and statistical characteristics (especially variability) of the process are defined, specifying sampling frequency becomes much more straightforward. Goals that involve estimates of average water quality over a given time period can be related to sampling frequency via levels of confidence in the estimate. Goals that involve detecting trends over time can be related to sampling frequency via power of the test for the trend, and goals that involve detecting compliance with standards can be related to sampling frequency via a percentile approach to compliance assessment.

While connections between information goals and sampling frequency can be made through statistics, there are a number of differing opinions as to whether such connections are correct or even appropriate, especially in a monitoring system that has a number of competing information goals. Calculations of sampling frequencies by a statistical method often results in optimization of sampling around one information purpose or goal. Yet, the system must attempt to meet several information goals simultaneously. It may, therefore, be best to evaluate logical alternative sampling frequencies in terms of their ability to satisfy multiple objectives, rather than to optimize for any one objective.

For example, it may be logical to sample at such a spacing in time that the samples can be assumed to be independent. This spacing in time would typically be approximately monthly in a stream situation and quarterly in a ground water situation (Loftis and Ward, 1980; and Montgomery et al. 1987). With such sampling, the data can then be evaluated as to their ability to generate the desired information in a manner such as illustrated by Hirsch et al. (1982). These authors, using total phosphorus data, compiled a national summary of stream-quality trends using the U.S. Geological Survey's National Stream Quality Accounting Network (NASQAN) data. They assessed trends by selecting and developing data analysis methods that meshed with the sampling frequencies and resulting data characteristics.

Regardless of the approach chosen to specify sampling frequencies, the key point of this discussion is to document the logic used in relating sampling frequency to the information to be produced. Such documentation is then included in the final monitoring system design report.

7.3 WATER QUALITY VARIABLES TO MEASURE

When the information goals of a water quality monitoring system are defined in a quantitative manner, the water quality variables on which information is sought are also defined. Such specification, however, does not imply that all such variables should be measured at the same frequency. There is often considerable information on a water quality variable that can be obtained by correlation to another variable or to several other variables (Sanders et al. 1983, Chapter 6). Thus, a major thrust of this portion of the network design is to translate the definition of "water quality" obtained from Step 1 (and the information expectation report) into a set of water quality variables that are to be measured routinely, along with a means of obtaining the desired information on other relevant variables measured less frequently.

The issue of determining which variables to measure in water quality monitoring is complex and rapidly evolving as the number of "new" compounds found in the environment increases, almost daily! Many recognize that it is impossible to measure everything in the environment and that some logical means of selecting variables to measure has to be a part of every water quality information system.

Systems focusing on long term trends in water quality will utilize more "indicator" type measurements, while those addressing a special investigation of a specific water quality problem will utilize the variables inherent to the problem (for example, those known to be spilled). Systems examining compliance with water quality standards will necessarily use the variables listed in the standards. Biological variables to measure could be directly specified in standards or in permits. More commonly, however, there are a number of biological indicators that could be selected to track ecosystem health in terms of either the functioning or structure of communities. It is often viewed as poor biology to focus only on one or two species. However, there is a persistent need to obtain biological information within time and budget limits. Cairns (1989) addresses many of these issues in an examination of the role of biological monitoring within the larger subject area of monitoring system design.

In almost all cases there is a perceived need to "monitor," or measure, more variables than can be justified from an information viewpoint. This perception is associated with a desire to "insure" against future findings that could create legal or regulatory problems. "Insurance" is difficult to incorporate into a monitoring system design, because the future quite probably brings science into the picture that did not exist at the time of the original monitoring system design. Consider how much our perception of what is harmful to human and environmental health has changed in the past ten years.

Budget realities make it impossible to "monitor everything" in hopes that this will provide the desired "insurance." This situation is leading into a search for easily measured indicators for large categories of variables. These indicators, if positive, suggest a need for more detailed laboratory analysis. If negative, then it can (safely?) be assumed that none of the larger group of chemical species is present.

Indicators are not new to water quality. *E. coli* has been used for years to indicate microbiological contamination. If *E. coli* are under control, then harmful microbiological organisms are assumed to be under control. Other measurements that can provide indications of water quality in a similar manner are now being sought, and, obviously, much more work is needed.

7.4 SUMMARY

Step 3 of the recommended approach to monitoring system design deals with the monitoring network itself, specifying the where, what and when of sampling. Many designs stopped at this point in the past. However, the overall systems view of monitoring requires that network design be derived from information objectives and that it directly supports data analysis, reporting, and ultimately information utilization. Thus, network design is only a small part of the total monitoring system design. The next chapter addresses the need to specify the ''rest of the system'' in an operational context.

8

SYSTEM DESIGN DOCUMENTATION

During the previous discussions of water quality monitoring system design it has been mentioned several times that the reasons for monitoring, the logic or rational behind network design, and the day-to-day operating procedures are often poorly documented, if documented at all. Other portions of most water quality monitoring systems, especially sampling and laboratory analysis, are generally well documented through the use of standard operation procedures or protocols. Such documentation, however, is not actually defined as part of a total information system design, but rather is documentation prepared by professional societies or government agencies as standards for general operation of a sampling or laboratory analysis effort. The connection of the sampling and laboratory analysis operating procedures to the information goals and objectives of the monitoring effort is missing.

This chapter describes a critical need for monitoring system design documentation, encompassing the entire water quality monitoring system, from preparing to sample to insuring that the information provided was actually utilized. Such documentation can be quite extensive and require considerable time and effort to produce. Once completed, however, the design document very clearly displays the results of the extensive design effort and may also serve as an educational tool directed toward those who are not familiar with the concept of monitoring as a complete information system.

This chapter deals with Steps 4 and 5 in the design process outlined in Chapter 4. These steps require documentation of all data collection procedures (Step 4) and all information generating and reporting procedures (Step 5). If Steps 1 and 2 have been completed in a thorough manner, a considerable amount of the documentation required by Step 5 will have already been completed.

Documentation of the design of water quality monitoring systems is a fairly new concept. It is very surprising to many people that an industry or government agency will spend millions of dollars monitoring water quality without a design of the system. Government would not attempt to conduct a census of the U.S. population without very careful design and documentation of the information sought and the means by which it will be obtained. An information gathering system that is not documented is open to question about any conclusions it generates.

In the design of a water quality information system, the rationale for selecting sites, sampling frequencies, variables being measured, sampling methodologies, laboratory procedures, data analysis graphics and statistics, and reporting formats influence the information produced. If the methods, procedures, statistics, and reporting formats for monitoring are developed after the initiation of sampling, the resulting data and ''information'' are subject to considerable uncertainty. The operators, having no documented procedures to follow, can make daily decisions on how they want to operate the system. Variability from operation of the system may then become large compared to the natural variability in water quality. At that point the data may yield little useful information.

8.1 MONITORING SYSTEM DESIGN DOCUMENTATION FORMAT

Perhaps the best way to explain the documentation of a water quality information system is through a sample ''Table of Contents'' for a design report, shown in the following:

Water Quality Information System Design Report

I. Introduction

  Water quality problem(s)
  - Definition of ''water quality'' to be managed

  Laws passed to address problem

  Management structure established to implement law

  Information needed to effectively operate management structure

  Purpose of design report
  - Describe design of monitoring system established to supply the needed water quality information

II. Quantifying Information Needs of Management and Expectations of Monitoring

Review of water quality management laws

- Information expectations implied by legal objectives
- Information expectations implied by management's functions and powers
- Directly stated monitoring requirements

Review of administrative structure employed to implement the law's management "tools" and their information needs

- Establishing criteria
- Setting standards
- Permitting procedures
- Enforcement mechanisms
- Construction loans
- Planning provisions
- Monitoring undertaken
- Reporting of water quality conditions

Information expectations placed on monitoring

- Definition of water quality "population" to be measured
- Behavior characteristics sought
- Statistical and graphical methods
- Form of statistical and graphical results
- Reporting format and frequency
- Information utilization framework

III. Confirming Chosen Statistical Methods

Review of data characteristics using past data

Evaluate ability of data characteristics to support assumptions of chosen statistical methods

IV. Network Design

Where to sample

What to measure

When to sample

V. Operating Procedures

Sampling

Laboratory analysis

Data handling

Data analysis

Reporting

Information utilization

VI. Implementation Schedule
   Time schedule of implementation
   Staff training
   QC for entire information system
   Evaluation of information system's performance
VII. References

The suggested framework for the final monitoring system design document utilizes much of the documentation already undertaken as part of performing Steps 1, 2, and 3 of the design process. Thus, the main additional work needed to prepare the final design document will focus on operational details more than why the monitoring is being performed.

In Section I of the final design document, it is important that the report spell out the water quality problems of concern, the laws passed to deal with these problems and the form of the management structure established to implement the law.

As the information needs are defined in Section II of the design report, the ways and means of producing the desired information must also be developed. With a realistic assessment of this ability, it may become apparent that some information sought simply cannot be produced with the current state of the art or the current monitoring budget! This portion of the design report must carefully document what information can be produced and rationalize this with the information sought.

The graphical/statistical means chosen to produce the information now agreed upon are identified in Section II of the design report. The purpose of Section III, then, is to confirm that the statistical methods chosen to support the conversion of data into the desired information do not have their assumptions violated to the point where the resulting information is incorrect.

Section IV is the place to document the location of sampling sites, sampling frequencies, and the variables to be measured, along with the reasoning behind these selections. Correlation between variables may reduce the number that must be routinely analyzed. Sampling frequencies suggested by the statistics must now be rationalized with the day-to-day operations of the total monitoring system.

Section V documents the actual operating procedures of the total system including sample collection, laboratory analysis, data handling, data analysis, reporting, and information utilization. Rather than repeat every detail in this section of the design document, standard protocols and procedures can be referenced.

Documentation of reporting procedures is especially critical because this is where the results or products of the entire monitoring system become visible. Documentation of the reporting procedures should include the following:

- Format of each report to be produced
- Frequency of publication of each report
- Audience to receive each report
- Distribution procedures for each report
- Content of each report (perhaps via example ''Table of Contents'' for each report)
- Types of conclusions to be drawn and presented in each report
- Automation of report preparation procedures using computer hardware and software
- Means for evaluating the effectiveness of each report

Simply preparing reports is not an appropriate end for an information system that is designed to support decision making within a water quality management program. There are too many examples of reports being prepared and shelved and their contents never entering the decision making process. Documentation of a water quality monitoring (information) system should follow the information to the point of its ultimate use and describe how this will take place. Such quantification can be threatening to decision makers; however, it must be pointed out very clearly that information on water quality is only one input into the decision making process and in no way is to serve as the sole basis for making decisions.

8.2 IMPLEMENTATION

Section VI of the design report, besides stating the timing for implementation, should address any remaining critical operational issues. These include, personnel, quality control/quality assurance, and assessment of the system's performance.

8.2.1 Personnel Mangement

Staff qualifications and training required to operate the monitoring system should be defined in the final design document along with a means of maintaining staff competence. The activities involved in water quality monitoring require a highly professional and well trained staff if accurate information is to be generated. Without a professional staff taking great pride in their work, the results of a well designed monitoring system will still be questionable.

How does a monitoring system manager acquire and retain a well qualified and up-to-date staff? This has been a problem with water quality monitoring system operations in the past, due in no small part to the rather vague role monitoring has played within a total water quality management effort. Too often new employees were assigned to the monitoring program until they had enough experience in the management effort so they could "move up to a better job." Such employee management practices places monitoring at the lowest end of the professional development ladder within an organization. These practices will result in considerable variation in water quality data due simply to changing personnel.

With respect to personnel operating a monitoring system, the ideal situation for generating accurate and reliable water quality information is to staff the monitoring effort with well qualified people who are encouraged to join professional organizations related to their job function. They should also be encouraged and supported to attend professional meetings and short courses in their field of expertise. There should be promotion levels within their monitoring specialization. They should be rewarded and recognized for their contributions to the operation of the monitoring system. Their names should appear on various "products" of the monitoring system related to their particular contribution. In other words, personnel working in a water quality monitoring system are professionals and should be treated as such. Their contribution to the success of the management effort is no less than that of any other employee, and the staff management system must be set up to recognize this fact.

8.2.2 Quality Control/Quality Assurance

A well designed and operated quality assurance/quality control (QA/QC) program is our only means of assurance that the data and information produced by monitoring are as accurate and precise as intended. The design document must define a first class QA/QC program including the staff time and resources required for operation. For larger monitoring systems, the best way to insure that QA/QC is properly implemented within the system is to establish a separate position responsible for all QA/QC efforts.

QA/QC is nothing more than the monitoring of the monitoring system to insure that it is performing as designed. Taylor and Stanley (1985) define quality assurance as:

> "... those operations and procedures which are undertaken to provide measurement data of stated quality with a stated probability of being right. The measurement system must be in a state of statistical control in order to justify such a probability statement. This is attained by quality control procedures which reduce and maintain random and systematic errors within tolerable limits, and by quality assessment procedures which monitor the quality control procedures and evaluate the quality of the data produced."

This definition, in the information context of this book, should also be expanded to include monitoring and supervision of the entire water quality information system and not just the data produced.

8.2.3 Reporting

In recognition of the importance of report preparation, the final design document should designate the individual or group charged with that responsibility and recognize the resource requirements of the task. Within water quality management programs there is an increasing recognition of the need to designate resources to the reporting function. Reilly (1989) states the case for better reporting and communication of environmental status to the public and their elected representatives. Portney (1988) notes that "we are woefully ignorant of the current state of our environment, of how the state compares with the past, and of the role the current policies may have played in accounting for the differences between past and present." To make matters worse, he says that existing data are not at all accessible to interested parties.

Portney (1988) suggests that one reason for this poor access to environmental information is the diffusion of effort among many different agencies and their many different monitoring systems. He then lists eight Federal agencies collecting environmental data. Ward (1989) lists 14 separate monitoring programs operated by the Virginia Water Control Board in their efforts to manage water quality. Even if all these monitoring efforts, within Federal, state or industrial levels, were well-funded and operated smoothly, it would still be necessary to gather the relevant data and sort through the mass of numbers to obtain the desired national, state, or industrial site environmental information (Portney, 1988).

Portney (1988) recommends that, at the Federal level, a Bureau of Environmental Statistics be established to serve the same role for the environment as: (1) the Bureau of Labor Statistics serves in the collection and reporting of data about unemployment, layoffs, and related matters; (2) the Bureau of Economic Analysis serves in supplying data on foreign trade, GNP growth and other economic measures; and (3) the Bureau of the Census serves in providing information on population growth, shifts, and other issues. He notes that all three bureaus "were created in part to ensure the independent and non-partisan treatment of data and measures that might be politically sensitive." Managing any country has, today, become dependent upon accurate information provided by such bureaus, and the environment is moving into that situation. Water quality reporting must be performed within the context of a total information system for each monitoring effort. And the organizational mechanism, be it a bureau or office or person, must be designated to accomplish the reporting function.

8.2.4 Evaluating the Monitoring System's Performance

To ensure that the monitoring system is continuing to serve the information needs of the managers, some provision must be made to evaluate the information product in terms of the design goals. Also, as information needs may change, provisions need to be made for the design of the system to evolve. It is best to anticipate that such changes will occur and to provide a means for addressing them in the initial monitoring system design document.

Assessing the information product of the monitoring system design can be accomplished in a number of ways. Informal enquires of information users, formal surveys of information users, and tear sheets (evaluation reports) in the reports sent out, are some examples of how to obtain feedback on the information being produced. Auditing the management/monitoring interface along with its need for, and use of, water quality information is another procedure that could be utilized. However such evaluation is to be accomplished, it should be defined in the monitoring system design document.

Invariably the types and forms of water quality information will change, and probably expand, over time. Surveying user needs, as noted above, is one way to be aware of this change. However, it is necessary to discuss in the initial design document how changes are to be accomplished. There should be no ad hoc tinkering with the monitoring system. All changes must follow a set procedure and then be documented as appendices to the initial design document.

Just as the initial design will be approved by information users or their representatives, any changes to the design must also be justified and approved. Consistent operation of the monitoring system over time is critical to obtaining information that is comparable. Ad hoc changes to monitoring systems have often reduced the value of historical data records, but with careful documentation of changes, the entire data record can maintain its value over time.

8.3 SUMMARY

At the end of the documentation phase of the monitoring system design process, a design report for the system is published. Preparation of a well documented water quality monitoring and information system design will not be accomplished in one afternoon, nor in several afternoons. However, a complete and high quality design report is a definite, substantive product generated as a result of spending significant time and money on the design exercise.

This design document will serve to guide operation of the entire monitoring system and will clearly connect the tasks of all employees, thereby serving to facilitate pride, responsibility and ''ownership'' in the entire information system. Such explanations will also help managers to understand the importance of each employee and, hopefully, will help prevent monitoring employees being reassigned whenever a ''crisis'' happens. Crises often break the smooth flow of a monitoring program's operation, and eventually the value of the information is reduced.

A well documented water quality monitoring and information system, tied closely to the decision making of a management program, also serves to better justify the money spent on the monitoring effort. As a result, future budget changes can be evaluated in terms of their anticipated effect on information produced and in the ability of management to make decisions based on that information.

9
CASE STUDIES

The previous eight chapters have presented the concept of designing a water quality monitoring system in its totality, focusing especially on the need to define information goals and the graphical and statistical means of obtaining the information sought. Examples and references have been provided during the presentations, but these do not bring out the strengths, weaknesses, and real world applicability of many of the concepts. The four case studies in this chapter are designed to fill this gap and illustrate the usefulness of the recommended design concept. Of necessity, these descriptions cannot explain in great detail all the work that has gone into each design and its implementation. However, they do illustrate the nature of water quality monitoring systems that have been designed in a thoughtful and quantitative manner.

The case studies were written by people directly involved in the design and operation of the monitoring systems. The emphasis and writing style, therefore, differ from one study to another and from the remainder of the text. The reference lists reported by the designers for three of the case studies also indicate the differences in information used in the design process. These differences are, in fact, reflective of the different approaches and emphases that can be utilized in performing and documenting the design of a water quality monitoring (information) system. As the authors suggested, considerable differences in implementation can be expected within the general guidelines set forth and within the limits imposed by an organized and scientific approach to design.

The first case study describes the groundwater quality monitoring system designed and operated by IBM's East Fishkill Facility in New York. This monitoring effort is over 10 years old and is constantly evolving along with the state-of-the-art in monitoring system design. The application, being at a corporate site, is highly focused in both its information production and its spatial coverage. It represents, however, a typical situation for many industries today. The case study also addresses the management setting within which monitoring must operate. Note, in particular, the interaction of the regulatory agency requirements and the direction of the monitoring system design. It should be pointed out that IBM had information objectives over and above those of the regulatory requirements. In particular, IBM was determined to know what was happening to groundwater quality over time and space in relation to their activities at the site.

The second case study is an overview of the U.S. Environmental Protection Agency's efforts to design a water quality monitoring system to assess the effects of changes in acidic deposition on surface waters in the United States. This project is referred to as the TIME Project (Temporally Integrated Monitoring of Ecosystems). The efforts taken to insure statistical correctness while providing clearly defined information is unusual in such a large scale, national water quality monitoring system. While all elements of this monitoring system have not been implemented, they illustrate the detail needed to design effective monitoring systems. The case study also describes the level of effort needed to coordinate the design of such a large scale monitoring program within a government agency. As with any national water quality monitoring system, many people, offices, and laboratories are involved, and coordination becomes critical.

The TIME monitoring system design reflects, more directly than many other design efforts, the "research" required to specify various components in the total water quality information system. Since the existing state-of-the-art was not adequate to design a monitoring system like TIME, U.S. EPA staff had to utilize special studies in such areas as statistics and data analysis, definition of the deposition processes, the role of biology in the monitoring effort, and the integration of all the data acquisition efforts into a total ecosystem concept. It is not unrealistic to envision that other design efforts will also involve "research" activities due to one or more poorly defined portions of the system, particularly in the area of data analysis, reporting, and information utilization.

The third case study describes the efforts of the Water Quality Centre in Hamilton, New Zealand, to design a national water quality monitoring system focused on stream water quality. The documentation of information goals, along with specification of the data analysis methods to achieve the goals, set the stage for quantitative determination of sampling frequencies, sampling locations, and variables to measure. The establishment of a database management system to support data storage, retrieval, analysis, and reporting represents a truly complete system design. The strength of this case study is that the various practical aspects of a monitoring system design are dealt with in a systematic and logical manner. In particular, the where, what and when of sampling (network design) are described quantitatively. The information goals are tied to a data analysis procedure and information expectations are realistically portrayed.

In the fourth case study, a water quality monitoring network is designed to detect trends in the acidification of lake waters in Quebec. The total monitoring system design is not discussed in the case study (ie there are no operating, data analysis, or reporting procedures stated). The paper strictly deals with the initial network design. In locating sampling sites, use is made of "classification" techniques to find groups of similar sites. Pielou (1984) and Legendre and Legendre (1983) are two references that explain the use and interpretation of classification techniques. Classification and ordination techniques have been used widely in ecological studies and are now finding their way into use in physical/chemical studies. Use of such techniques, in conjunction with existing data from an area, permit much more refined monitoring system designs focused directly on the information needed for management.

9.1 IBM GROUNDWATER QUALITY MONITORING PROGRAM AT EAST FISHKILL, NEW YORK

(Prepared by Harry F. Bell, Senior Scientist, IBM Corporation, Hopewell Junction, New York)

IBM's East Fishkill plant is one of the world's largest semiconductor manufacturing facilities. Approximately 10,000 employees and 4 million square feet of buildings occupy 750 acres in southeastern New York. Large volumes of inorganic and organic chemicals are used in the manufacturing process. The plant began operations in 1963.

The plant is located in a semi-rural area and there are no municipal water supply or wastewater treatment facilities nearby. The site is totally dependent upon groundwater resources for process and potable water use. The plant currently uses approximately 2 million gallons per day (MGD). The water is withdrawn from a limestone aquifer, used for a variety of plant uses, treated and ultimately discharged to surface streams at points permitted under the New York state discharge permit system.

The site is surrounded on three sides by homes and a high school, all of which are dependent on groundwater supplies (see Figure 9.1). The private homes dispose of their wastewater via septic systems.

As part of IBM corporate environmental policy to ". . . meet or exceed all environmental regulations" the facility began a series of

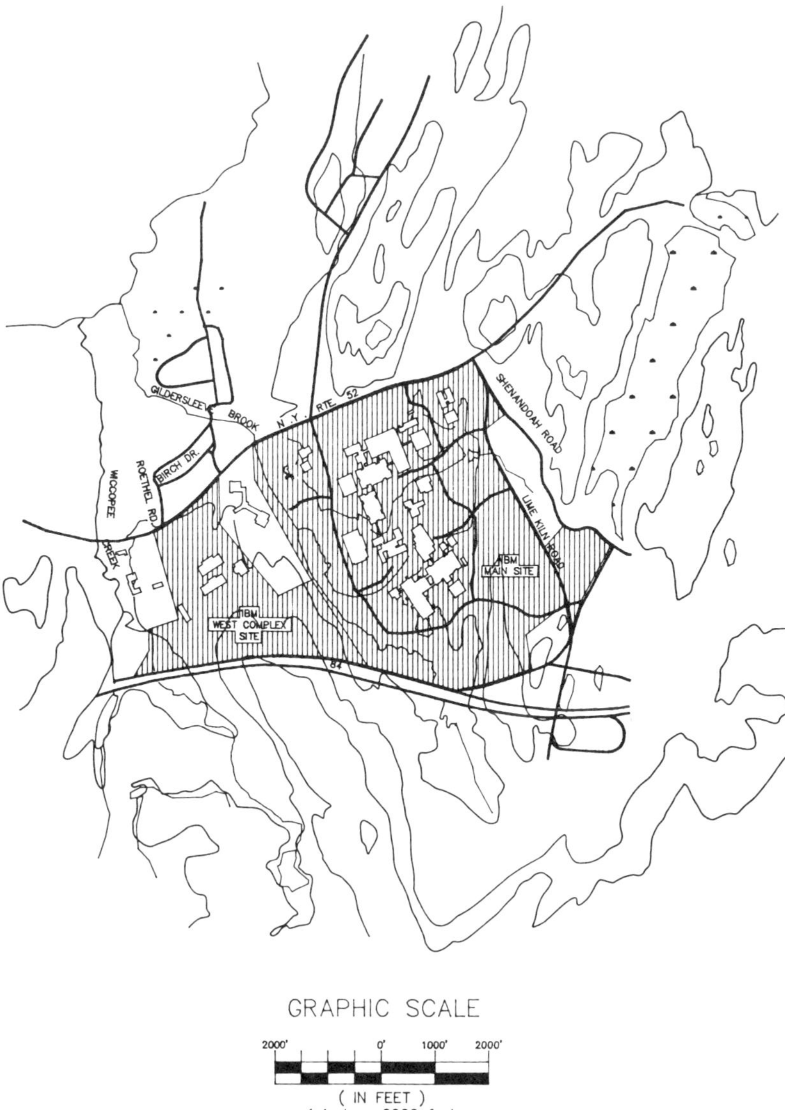

Figure 9.1 IBM site locations, East Fishill, New York.

groundwater tests in 1978. This initial investigation identified four areas on the site with groundwater concerns (see Figure 9.2). The program involved a review of the known hydrogeology of the site and the installation of approximately 65 monitoring wells. Water quality data (for inorganic and organic compounds) were collected from the monitoring wells and from the site's production wells.

Additional investigative work demonstrated that the problem was primarily related to the presence of chlorinated hydrocarbons (*Di, Tri,* and *Tetra* chloroethenes with 1,1,2,2-tetrachloroethene representing the most commonly occurring compound). Levels of the solvents ranged

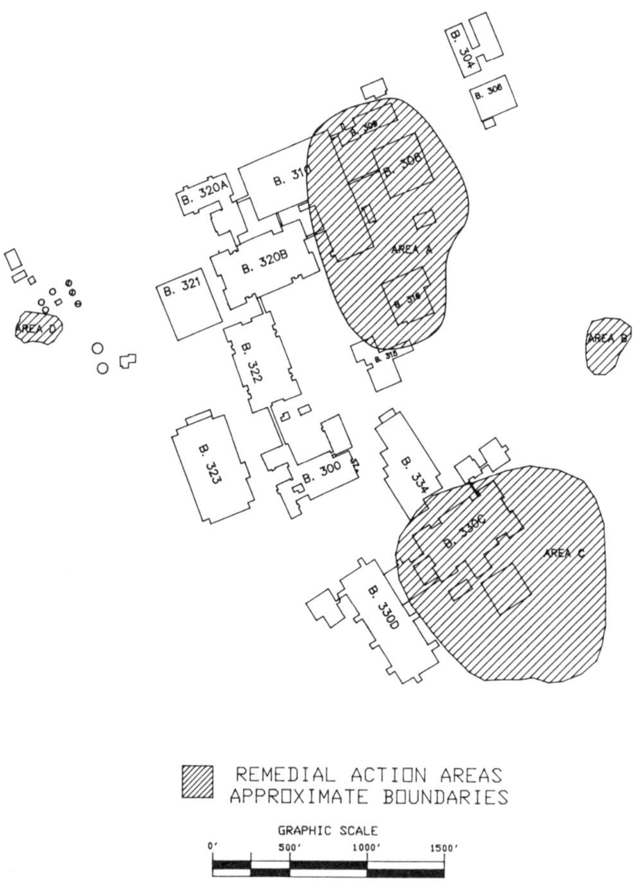

Figure 9.2 Remedial action areas, approximate boundaries.

from non-detect (<1 μg/L [ppb]) to separate phase materials in glacial overburden soils and non-detect to 10 mg/L [ppm] levels in the fractured (Dolomite) bedrock. Chlorinated hydrocarbons were also detected in production wells in the general vicinity of the contaminated areas.

The source(s) of solvents were identified as leaks and spills associated with solvent storage and handling areas, two former fire brigade training areas and a former construction debris landfill. No major tank failures were found. These spills were assumed to have occurred over the time period of 1963–1978. It was at this time the work began to replace all buried tanks with above ground, double walled tanks located in catch basins. Many subsurface pipelines have been replaced with pipes (with secondary containment) on trestles.

With this discovery of chemicals in the groundwater, a groundwater quality monitoring program was begun in 1979, using the design methods of that time period. Existing wells were augmented with new wells drilled at points considered critical to developing a better understanding of the groundwater quality situation on-site. Samples were taken at regular intervals, analyzed in a laboratory, and the results were computerized.

In addition to the on-site groundwater investigations and routine monitoring, two off-site water quality programs have been performed (1981–1982 and 1986–1988) in the area immediately adjacent to the IBM facility. Both of these programs have demonstrated that no IBM related organic chemicals are in private water supplies. The large volumes of water pumped by IBM to meet its local demand had created a cone of depression under the site that had groundwater flowing toward the IBM site.

Remediation efforts are being carried out on the four known contamination areas under an Order-on-Consent signed by IBM and the New York State Department of Environmental Conservation (NYSDEC) in 1981.

The Order-on-Consent specifies that IBM will:

1. Contain, collect, and treat contaminated groundwater.
2. Implement remedial action programs.
3. Establish a groundwater monitoring program.

The order further specifies that IBM will notify NYSDEC if the contaminated groundwater has negative impacts on off-site water supplies, IBM's production wells or if the remediation efforts are experiencing difficulties.

To address the first two directives, IBM established a treatment system that pumps and then treats (with air stripping and/or granular activated carbon) approximately one million gallons a day of contaminated groundwater. More than 75% of this water is used (after treatment) for process and potable purposes. IBM also completely removed 20,000 cubic yards of the old construction landfill and disposed of it in a secure hazardous waste depository. The area was refilled with clean soils, capped with clay and the underlying groundwater is pumped and treated by a production well.

The purpose of this case study is to describe IBM's evolving efforts to design and operate a state-of-the-art groundwater monitoring program and to describe the "information" obtained. These are the efforts that have been undertaken to meet the third requirement of the NYSDEC Order-on-Consent.

9.1.1 Initial Groundwater Monitoring System Design (1982)

In 1982 IBM undertook the documented design of a total groundwater quality monitoring system (the 1979 design was not documented). This effort first involved characterization of existing water quality data. There were, by this time, considerable data (3400 samples had been processed from 290 sampling points). The sampling frequency was not consistent and there were many missing values in the data record. The data, when plotted against time, revealed considerable variability. In addition, there were a large number of non-detects in the data record. The conclusion was that there was actually little information in the data and that it was not very useful for design purposes.

The monitoring system design proceeded with the following being defined:

1. Variables to be measured were mainly those associated with IBM's prior actions at the site (chlorinated hydrocarbons with emphasis on the four common compounds).
2. Sampling locations were grouped by site perimeter, contaminated areas and general coverage (and related to the information needed to assure that the groundwater quality objectives were being met).
3. Sampling frequency was based on the importance of the locations of the wells and past water quality variability.

At the same time, a data verification protocol was implemented. This involved comparing new data with ranges of past results (see Figures 9.3 and 9.4). A data management system was established using an IBM personal computer and a commercially available database software pack-

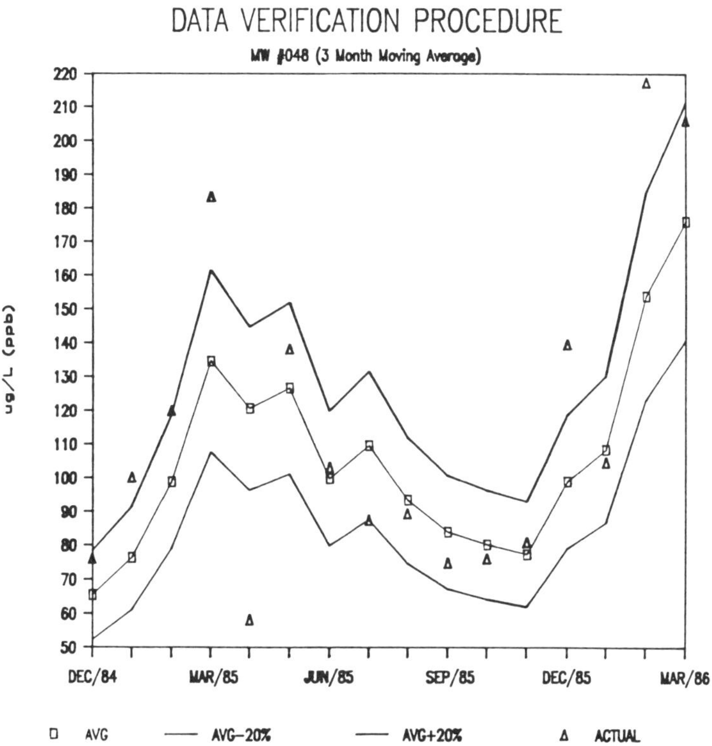

Figure 9.3 Plot illustrating the data verification procedure.

age (see Figure 9.5). Data analysis methods were identified to provide the information deemed necessary at that time (see Figure 9.5). Reporting mechanisms were defined to include rather graphical means of presenting the results to managers (see Figure 9.6 for a list of the reports to be produced and Figure 9.7 for examples of the contents).

The initial design also called for regular evaluations of the data to determine if the original assumptions underlying the design were valid. As higher quality data were obtained, it would be possible to evaluate the statistical behavior of the variables being measured and ensure that the information being generated was statistically sound.

9.1.2 Groundwater Monitoring System Today

The groundwater monitoring system today includes over 470 wells and piezometers, some of which are over two miles from the main facility.

DATA VERIFICATION PROTOCOL

- COLLECT LEAST CONTAMINATED FIRST
- FIELD SPLIT SAMPLES
- ANALYZE ONE SPLIT

  COMPARE WITH AVERAGE OF LAST THREE
  ACCEPT/RECORD IF WITHIN PLUS/MINUS 20%

  OUT OF RANGE, REVIEW DATA/REANALYZE/ RESAMPLE

- NEW SAMPLE/REPLICATE IN RANGE, RECORD

  STILL OUT OF RANGE, ASSUME WATER QUALITY CHANGE, RECORD

Figure 9.4 Steps in the data verification procedure.

Approximately 100 of the wells are routinely sampled for water quality. Many of the piezometers are used for elevation measurements related to the production wells' operation.

The monitoring program today consists of the collection, analysis, and reporting of both routine and non-routine water quality data. The means of dealing with non-routine monitoring requirements had to be established so as not to interfere with the routine monitoring (it was the "crises" of earlier management efforts that had resulted in missing data in the earlier data records). In addition to water quality monitoring, the system today records groundwater elevation measurements at all monitoring wells and piezometers (both on- and off-site).

All sampling and analysis is carried out by IBM personnel at the East Fishkill facility. More than 2000 routine samples are analyzed yearly for volatile organic compounds. Less frequent samples are taken for priority pollutant, inorganic, radiological and bacteriological analysis.

Since IBM East Fishkill is a non-community water supplier, many routine samples are taken throughout the supply and distribution network. Emergency procedures are in place to handle any contingencies which

DATA BASE FILES
(PC & DBMS SOFTWARE)

- WATER QUALITY

  WELL ID, SAMPLE DATE, PARAMETER CONCENTRATIONS

- ELEVATIONS

  WELL ID, SAMPLE DATE, ELEVATION (AMSL)

- PHYSICAL WELL DATA

  WELL ID, LOCATION, COMPLETION DETAILS, SAMPLING PROTOCOL, COMMENTS

DATA ANALYSIS

- PARAMETRIC ANALYSIS

  PRIMARILY CONTAMINATION AREAS

  * TIME PLOTS
  * DATA DISTRIBUTIONS
  * TWO SAMPLE t-TESTS (MEANS)
  * SIGNIFICANCE OF LINEAR FITS (TRENDS)

- NON-PARAMETRIC ANALYSIS

  NON-CONTAMINATED AREAS

  * HISTOGRAMS OF WATER QUALITY (VISUAL ANALYSIS), OUTLIERS

Figure 9.5 Data base management and data analysis specifications.

INFORMATION REPORTING

- MONTHLY BRIEFING
  - EXECUTIVE MANAGEMENT
  - GENERAL STATUS
- QUARTERLY REPORTS
  - TECHNICAL PERSONNEL
  - TECHNICAL STATUS
  - PROGRAM CHANGES
- ANNUAL REPORT
  - MANAGEMENT/TECHNICAL PERSONNEL
  - INPUT FOR EXTERNAL REPORTS

Figure 9.6 Reports to be prepared using ground water quality data.

might arise in the distribution system so that alternative sources of water could be provided for process and potable use.

Today, the overall goal of the monitoring program is the development of an understanding of the groundwater behavior beneath the East Fishkill site. This understanding is achieved through the collection of water quality and elevation data which is then integrated with the geologic and hydrogeologic data, analyzed and reported to the various requestors as an information product. More specific objectives of the monitoring system, with respect to water quality, are presented in Figure 9.8.

9.1.3 Data/Information Evaluations

In recent years the data records, containing much more accurate water quality data, have been developed and have created the ability to carefully examine the behavior of water quality at the East Fishkill facility. For example, Figure 9.9 is a time plot for monitoring well No.053 which clearly shows the effects of the monitoring system redesign which was

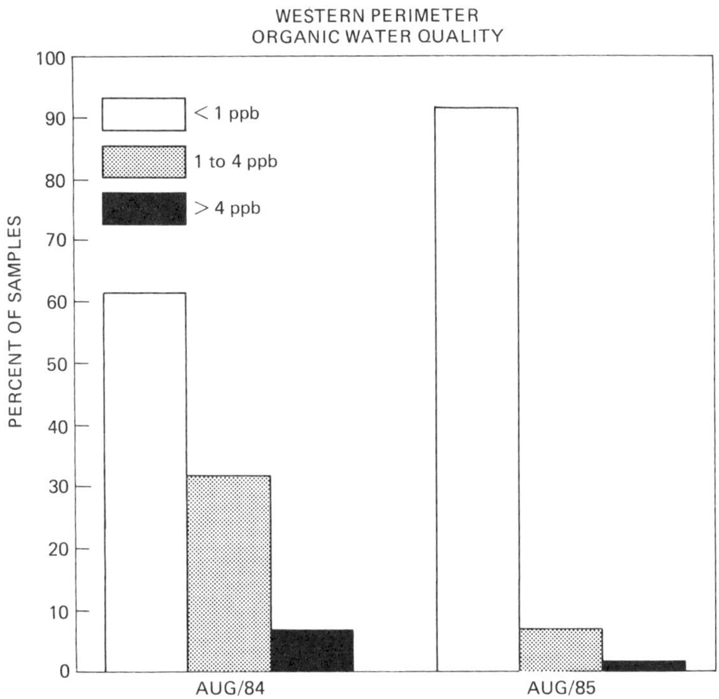

Figure 9.7 Examples of information reporting formats.

implemented in July 1983. Prior to this date the data contained considerable variability which has been attributed to the operation of the monitoring system (sampling and analytical errors) and not the behavior of groundwater quality. After the date, the data contain much less variation and represent the water quality behavior better. Although some remediation efforts were carried out in the vicinity of this well, the observed time dependent pattern of data variability reduction is common in many of the monitoring wells spread over a wide geographic area. It may therefore be concluded that the reduction in data variability is primarily the result of the monitoring system redesign.

With the data now representing water quality variation better, the earlier statistics were reevaluated. The presence of non-detects in the data record created a large barrier to better data analysis and development

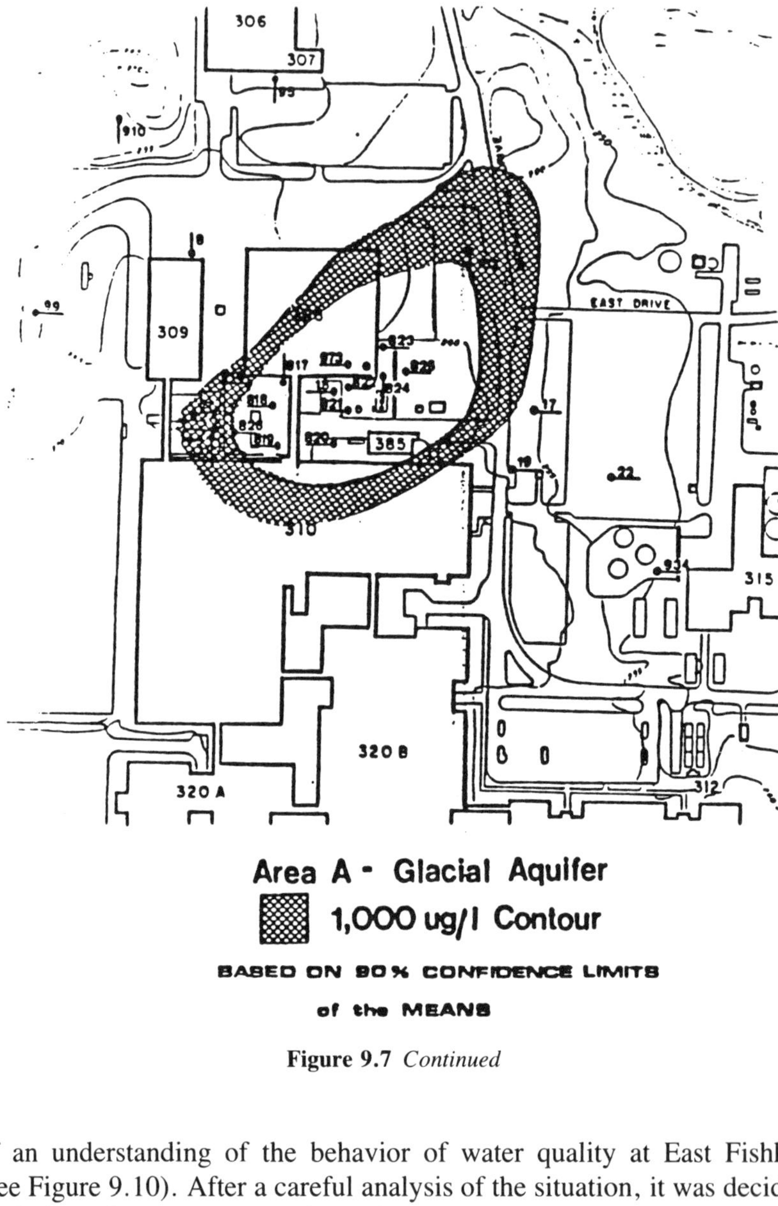

Figure 9.7 *Continued*

of an understanding of the behavior of water quality at East Fishkill (see Figure 9.10). After a careful analysis of the situation, it was decided to follow the recommendation of Porter (1986) and ASTM Method D4210-83 and not censor the data from the laboratory. The laboratory was asked to report all the readings along with an estimate of the error in the measurements. The uncensored laboratory measurements are used for all data calculations and then the results are rounded to the appropriate number of significant figures prior to reporting. Standard procedures

BASIC MONITORING OBJECTIVES

* DETERMINATION OF AVERAGE CONDITIONS
  - DIFFERENCES AS A FUNCTION OF TIME & SPACE
* DETERMINATION OF TRENDS
  - BEHAVIOR OVER TIME
* DETERMINATION OF COMPLIANCE WITH REGULATIONS
  - TRENDS IN COMPLIANCE OVER TIME

Figure 9.8 Water quality monitoring objectives.

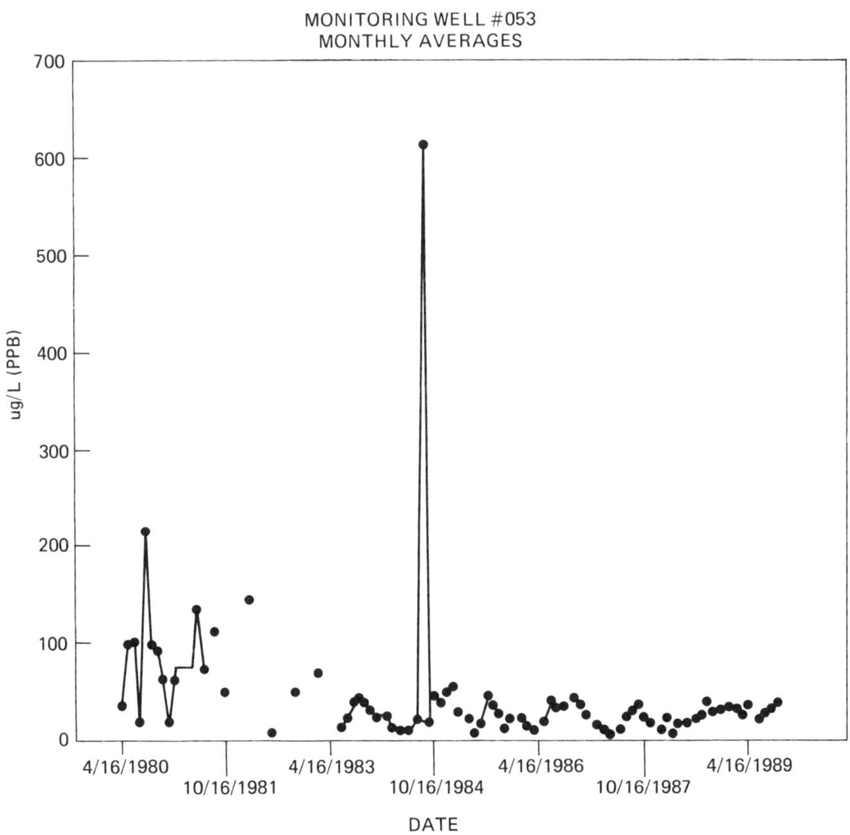

Figure 9.9 Time plot of tetracholorethene at well No. 53.

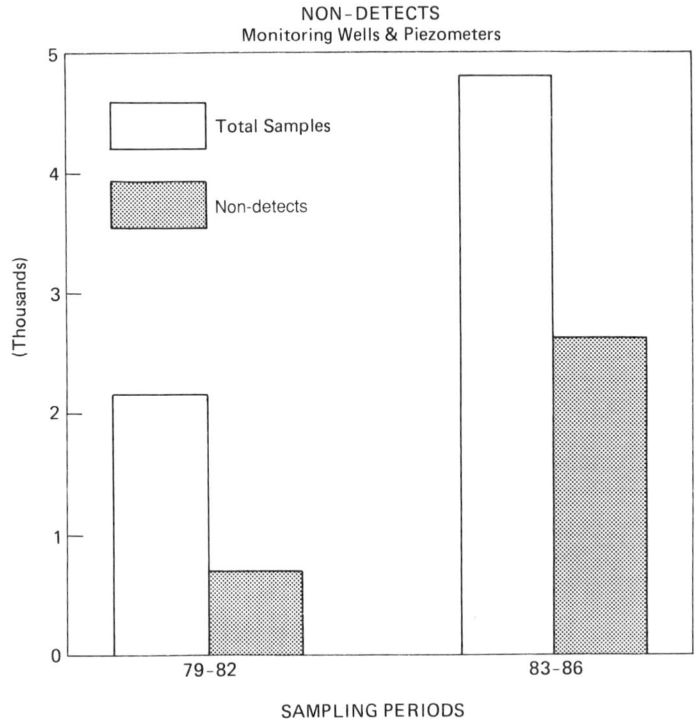

Figure 9.10 Number of non-detects in the data record.

for the rounding of data are available in many of the analytical methods employed by the laboratory. While initially difficult to implement, the results have been very helpful in supporting the development of information from the monitoring system. Figure 9.11 illustrates the advantages of working with an uncensored data set. Uncensored data were collected and analyzed for the well (Figure 9.11) throughout the time period of interest. The method detection limits (which changed during the course of the study) are shown on the figure. The additional information gained by the collection and analysis of the uncensored data resulted in an early warning and installation of a treatment system long before it was required.

During this period of evaluating the monitoring system's data analysis procedures, it was realized that data analysis is very much a personal decision with little opportunity for anyone to audit the process. Following

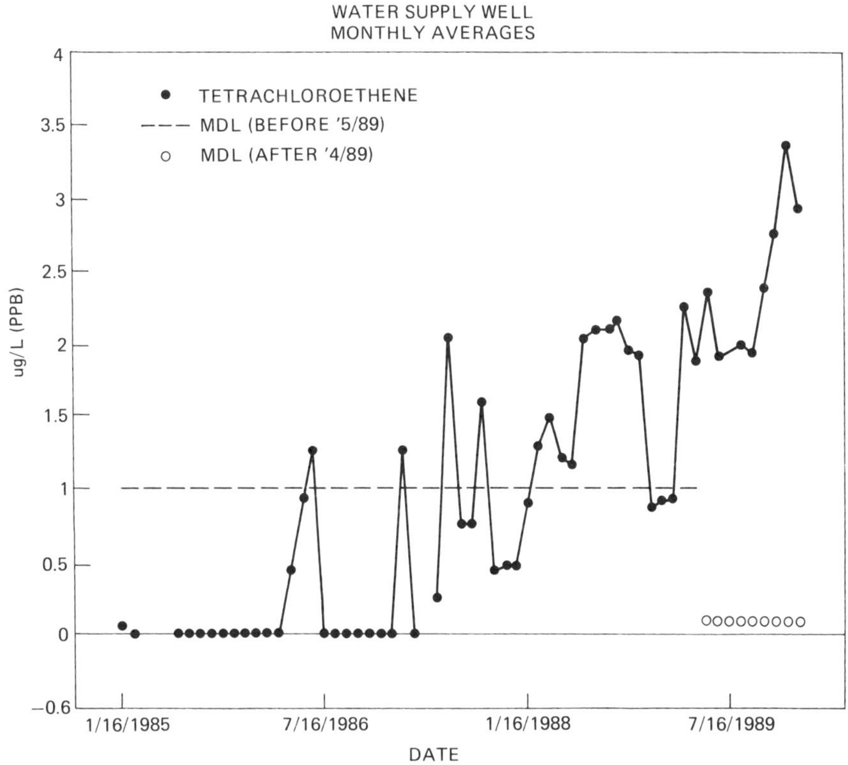

Figure 9.11 Time plot with changing detection limits.

the concepts of quality assurance that had guided the evolution of the monitoring system up to this point, it was thought that the operation of the "information" end of the monitoring system needed to be carefully defined in a protocol in much the same way as the means used to collect and analyze samples had been. This led to a careful evaluation of the concept of a groundwater data analysis protocols and to the proposing of a protocol that could be used as part of groundwater quality monitoring systems (Ward et al. 1988).

The data analysis protocol has to account for a number of data characteristics that can greatly influence the results of data analysis (See Figure 9.12). To illustrate, the data from well No. 48 have been analyzed for seasonality, trend, normality, and serial correlation. The results are pre-

DATA CHARACTERISTICS

* CONTROLLABLE CHARACTERISTICS
  - MISSING DATA (CONSISTENT SAMPLING PROGRAMS)
  - NON-DETECTS (CENSORED DATA SETS)
  - HIGH VARIABILITY (SAMPLING & ANALYSIS ERRORS)
* UNCONTROLLABLE CHARACTERISTICS
  - SEASONALITY
  - TRENDS
  - DATA DISTRIBUTIONS (NORMAL vs NON-NORMAL)
  - SERIAL CORRELATION (INDEPENDENCE OF SAMPLES)
  - HIGH VARIABILITY (ASSOCIATED WITH GROUNDWATER FLOW and CONTAMINANT TRANSPORT)

Figure 9.12 Data characteristics needed to be addressed.

sented in Figure 9.13. As the data record lengthens, the conclusions to be drawn from the data change. With too little data, it is difficult to obtain conclusive information. These factors have to be considered and built into the data analysis protocol. Work continues today on a data analysis protocol that incorporates all these factors in a statistically correct manner.

9.1.4 Groundwater Program Management

Groundwater monitoring at East Fishkill is designed to support groundwater management efforts. Groundwater protection is a long term process which must be recognized in the design of monitoring and management efforts. The normally slow movement of groundwater, the transport of a contaminant from the surface through an unsaturated zone to a saturated zone, the geologic-hydrogeologic conditions and the effects of retardation, biodegradation and dispersion all combine to make the understanding and remediation of groundwater problems extremely difficult and time consuming. Considerable periods of time may be required before a problem may be identified, the scope of the problem understood, and action taken. During and after remediation efforts, monitoring must be carried out for long periods of time to ascertain the success of such projects.

While the current water quality monitoring system is operating to

MW #048 CHARACTERIZATION SUMMARY

DATA SET SIZE MONTHS	SEASONALITY	TREND	NORMALITY	SERIAL CORRELATION
9	NO	NO	NO	NO
21	NO	NO	NO	NO
33	NO	YES	NO	NO
45	YES	YES	NO	NO
57	YES	YES	YES	NO
69	YES	YES	YES	YES
81	YES	YES	YES	YES
90	YES	YES	YES	YES

ALL TESTS AT THE 90% CONFIDENCE LEVEL

Figure 9.13 Statistical evaluation of data from well No. 48.

provide the information needed to support management objectives, its success in accomplishing this job is constantly being questioned. This on-going questioning is what has led to improvements over the years. Current efforts are focused on the development of data analysis protocols that supply the desired information using scientifically correct methodologies. Efforts are increasing to better understand the role that spatial correlation may play in the interpretation of groundwater monitoring data. A more precise knowledge of the role of data characteristics (such as spatial and temporal correlation) will assure that data analysis protocols are correctly applied. The integration of physical and statistical groundwater data will yield a more comprehensive understanding of the behavior of water quality at the East Fishkill site.

9.1.5 Summary and Conclusions

The groundwater monitoring program at IBMs East Fishkill has evolved with the technology employed to operate it. Figure 9.14 is a summary

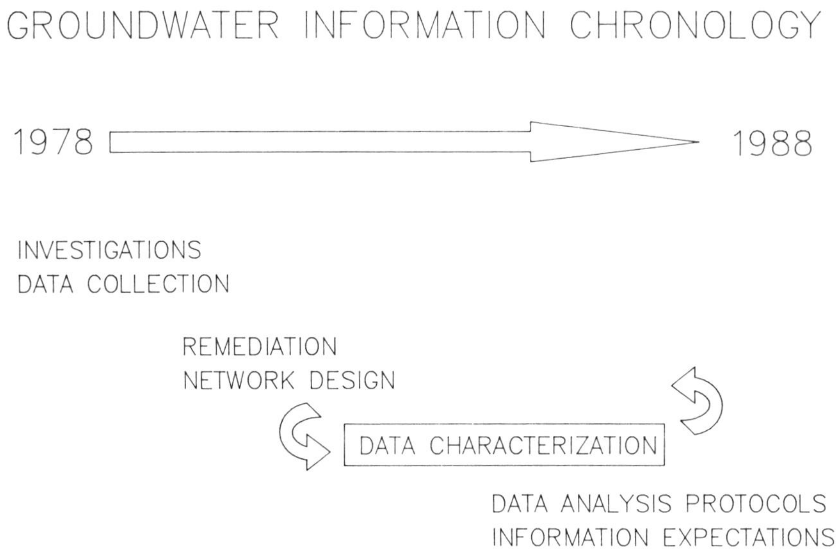

Figure 9.14 Summary of ground water monitoring system evolution at IBMs East Fishkill Facility.

of this evolution. It is not complete as the task of groundwater quality management, and the monitoring employed to support it, has not reached a mature state-of-the-art. As high quality data records lengthen, it becomes possible to use additional analysis methodologies on the data to attempt to obtain a better understanding of the behavior of water quality and the system designed to measure it.

IBM East Fishkill has had a strong focus within its water quality management and monitoring efforts to understand the importance of the statistical characteristics of data sets and how this information can be supportive of management decision making. The combination of information objectives and a data analysis protocol results in a system which provides the ultimate user of the data with a useful information product in an environmentally sound and cost effective manner.

REFERENCES

Porter, P.S. 1986. Statistical analysis of water quality data affected by limits of detection. PhD Dissertation, Agricultural and Chemical Engineering Department, Colorado State University, Fort Collins, Colorado 80523

Ward, R.C., J.C. Loftis, H.P. DeLong, and H.F. Bell. 1988. Ground water quality: a data analysis protocol. Journal Water Pollution Control Federation, Vol. 60(11):1938–1945, November.

9.2 TIME SYSTEM DESIGN

(Prepared by Allison K. Pollack, Systems Applications, Inc., San Rafael, California, and Jesse Ford, U.S. Environmental Protection Agency, Environmental Research Laboratory, Corvallis, Oregon).

This case study describes the U.S. EPA's monitoring project to assess the effects of changes in acidic deposition on surface waters of the United States. This project is referred to as TIME (Temporally Integrated Monitoring of Ecosystems).

The TIME Project has been in the conceptual and design stage during the period 1987–1989. This planning period has allowed careful attention to many aspects of network design not usually covered prior to field implementation. The network design issues discussed in this paper are addressed within the context of the need for a policy-relevant, cost-effective national monitoring network to assess the effects of changes in atmospheric deposition of specific pollutants (sulfate and nitrate) on specific and well defined target resources (U.S. surface waters of low acid-neutralizing capacity (ANC)).

Although the TIME Project was originally intended as a stand-alone monitoring program as part of EPA's activities under the National Acid Precipitation Assessment Program, it was incorporated in 1989 into the larger Environmental Monitoring and Assessment Program (EMAP) being planned by EPA. EMAP is designed to assess and periodically document the condition of all ecological resources (e.g., forests, wetlands, surface waters) of the United States (EPA 1990). TIME, on the other hand, has been designed to assess the effects of changes in loadings of one particular pollutant type on one subset of a single resource type. These complementary activities can both be achieved within the EMAP structure, and many of the planning activities that have gone into TIME network design now contribute to EMAP-Surface Waters design.

In this case study, we first discuss the 1987–1989 planning goals of the TIME project. The authors describe previous EPA projects relating to aquatic effects of acidic deposition upon which the TIME design is based. In the third section we describe a conceptual model for TIME that has helped focus planning activities. The last section describes the purpose and elements of TIMEs Data Analysis Plan, which will now become part of EMAP-Surface Water's Data Analysis Plan. This paper, in describing an evolving monitoring system design, updates information recently published in a closely related paper (Pollack and Ford 1989) on the same subject.

9.2.1 Background and Project Goals

Oxidation products of atmospheric emissions of sulfur and nitrogen oxides constitute the main components of acid rain. Anthropogenic emissions of one of these components, sulfur dioxide, have been declining since the early 1970s in the northeastern United States (Husar, 1986). However, little evidence is available to suggest that the rate of lake and stream acidification in these areas has changed (Driscoll et al. 1989; Schindler, 1988). Indeed, although sulfur emissions have declined, nitrogen oxide emissions and deposition have increased, and some studies indicate that the net result may have been little change in total rainfall acidity (Hedin et al. 1987).

Recent proposals for reducing acidic deposition have targeted sulfur dioxide emissions and call for reductions of 5 million tons by 1995 and for an additional 5 million tons (plus 2 million tons of nitrogen oxide emissions) by the year 2000. These reductions will leave sulfur dioxide emissions at approximately 60 percent of current emissions. Given the magnitude of investment required to achieve this reduction, it is appropriate that a well-designed long-term monitoring program be in place to assess the effectiveness of this decrease in emissions on aquatic resources currently affected and at risk of further degradation.

The 1987–1989 planning goals of the TIME project have been to:

- provide early-warning signals of surface water acidification (or recovery) in regions of interest,
- provide an ongoing assessment of regional patterns or trends in surface water acidification (or recovery),
- assess the relationships between observed patterns and trends in surface water quality and observed patterns and trends in atmospheric deposition, and
- assess the extent to which observed patterns and trends in surface water chemistry correspond with model forecasts of surface water chemistry.

TIME has been designed to collect long-term data on acid-sensitive lakes and streams in areas of the United States considered potentially susceptible to the effects of acidic deposition (Thornton et al. 1987; Payne and Ford, 1988). The project is designed to use information on wet deposition available through existing monitoring networks that may be augmented to meet project goals. As of March 1990, the project was scheduled for implementation in 1991.

9.2.2 Related EPA Studies on the Aquatic Effects of Acidic Deposition

The TIME project focuses on regions of the United States where there are lakes and streams with low levels of acid neutralizing capacity (ANC). These sensitive regions have been identified and studied in two types of major projects now underway or recently completed by EPAs Aquatic Effects Research Program (AERP). The first type includes large-scale surveys designed to identify the status and extent of surface water acidification in the United States. The second focuses on more detailed characterization and process-oriented research on selected systems, in an effort to provide an understanding of the underlying mechanisms responsible for the regionally observed patterns and effects. We briefly describe here those surveys and projects most closely related to, and supportive of, the TIME project design.

The regional studies are collectively known as the National Surface Water Survey (NSWS). The NSWS consists of two main components, the National Lake Survey (NLS) and the National Stream Survey (NSS). The NLS in turn consists of two major regional surveys, the Eastern Lake Survey (ELS; Linthurst et al. 1986 and Landers et al. 1988) and the Western Lake Survey (WLS; Landers et al. 1987 and Eilers et al. 1988). The purpose of these regional synoptic surveys, all of which have now been completed, was to provide a snapshot characterization of the populations of interest. The regions surveyed in each of these projects are shown in Figure 9.15.

The Eastern Lake Survey was conducted in two phases. In Phase I, 1798 lakes were sampled once during the fall turnover period in 1984 in the Northeast, Upper Midwest, and Southeast. Phase II of the survey was conducted in 1986; 145 of the lakes sampled in the Northeast in Phase I were sampled once in each of the spring, summer, and fall seasons. During the Western Lake Survey, 719 lakes in the mountainous areas of the western United States were sampled during fall turnover in 1985. The WLS was recently augmented by the Kenai Lakes Investigation Project (KLIP), during which 59 lakes on the Kenai Peninsula of Alaska

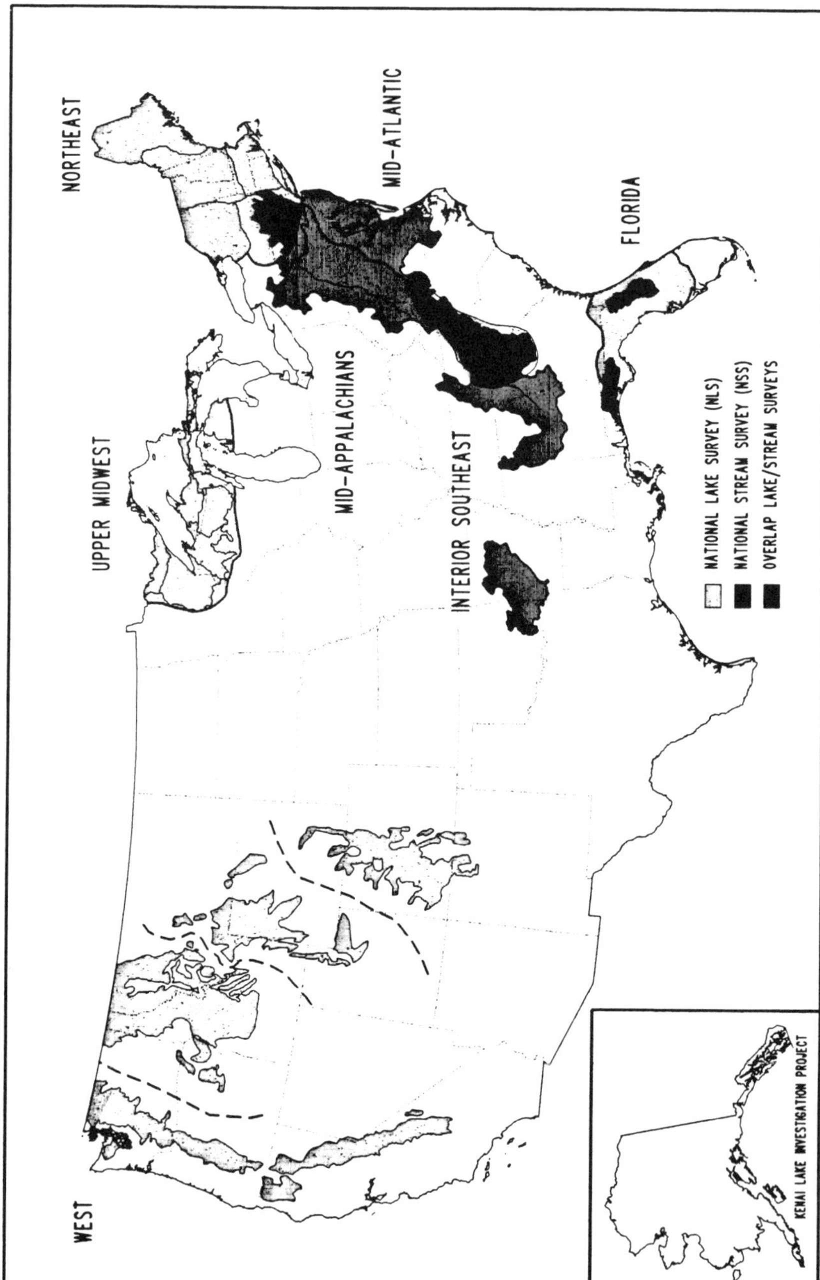

Figure 9.15 Regions of interest in the National Surface Water Survey.

were sampled during August 1988 (Ford et al. 1988; Landers et al. 1989).

The NSS also was conducted in two phases. In the first, pilot, phase 54 stream reaches were sampled in the spring and summer of 1985 in the Southern Blue Ridge to assess the logistic and scientific feasibility of a full-scale survey (Messer et al. 1986; 1988). The full survey was then conducted in the Mid-Atlantic and Southeast in the spring of 1986, when upstream and downstream reach ends were sampled at 445 stream reaches (Kaufmann et al. 1988).

The NSWS surveys were designed to be statistically representative subsets of the population of resources at risk. For the NLS, the lake population was divided into strata defined on the basis of region, subregion, and preliminary versions of mapped alkalinity class (Omernik and Powers, 1983; Omernik and Kinney, 1985; Omernik and Griffith, 1986a, 1986b), and a probability sample of lakes (specifically, a systematically sampled list frame of lakes with a random start) was drawn within each stratum. The NSS also was stratified by region and subregion, but because listing the extensive number of streams was not feasible, a two-stage sampling scheme based on a point frame within strata was employed. Thus, in both the NLS and the NSS, systems were selected for sampling within subregions with known, but unequal, probabilities (Blick et al. 1987). This design allowed for determination of the percentage (by number and area), location, and chemical characteristics of surface water populations in potentially sensitive regions and subregions. The probability sampling approach also provides both context and rationale for selecting lakes and streams for longer term monitoring for acidic deposition effects.

One of the objectives of the NSWS was to determine the proportion of lakes or streams having a certain chemical value less than or equal to a specified value. Such a proportion is easily estimated as the ratio of the estimated number of lakes (or streams) that meets the criterion divided by the estimated number of lakes (or streams) in the population of interest. This ratio can be calculated for the whole range of criteria encompassed by the data, and can be plotted as a relative cumulative frequency distribution function (Sanders et al. 1983); curves can then be overlain to compare characteristics of regional or subregional populations. Figures 9.16*a* and 9.16*b* are examples of these curves for ANC in lakes in all NLS regions and in WLS subregions, respectively. Median ANC values can be compared by extending a horizontal line from 0.5 on the ordinate; thus in Figure 9.16*a* we see that median ANC is lowest in Florida (about 75 μeq/L) and highest in the Upper Midwest (about

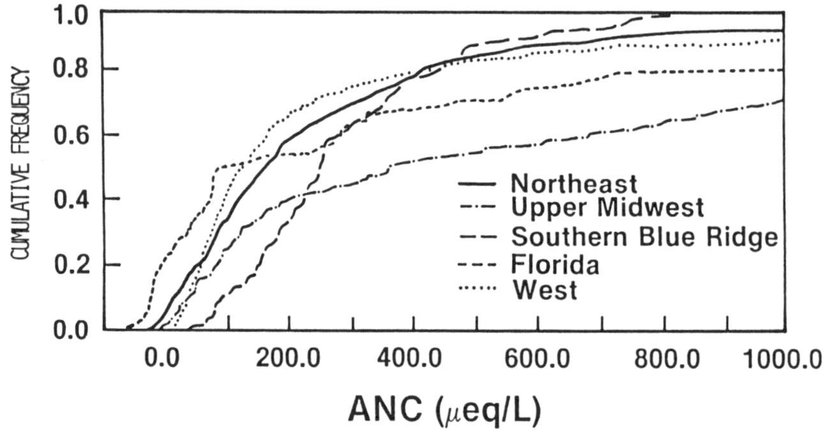

Figure 9.16*a* Cumulative frequency distributions for acid neutralizing capacity in National Surface Water Survey regions.

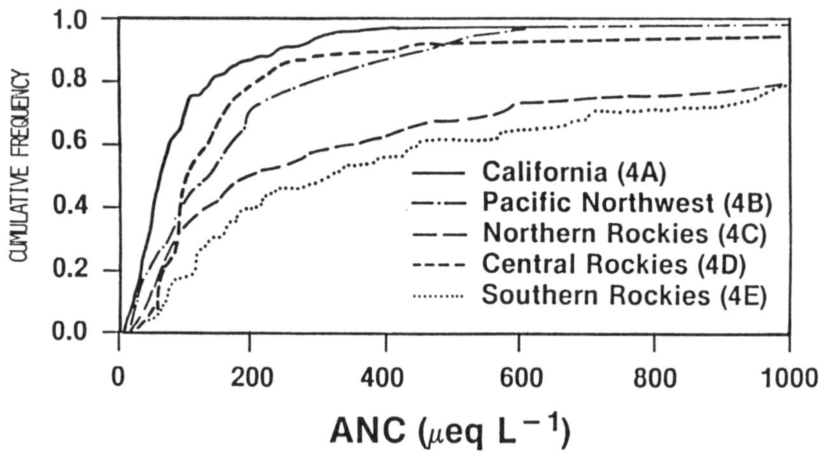

Figure 9.16*b* Cumulative frequency distribution for acid neutralizing capacity in lakes in subregions of the Western Lake Survey (from Landers et al. 1989).

350 μq/L). The proportion of lakes with ANC less than 200 μeq/L can be compared by extending a vertical line from 200 on the abscissa; in Figure 9.16*b* for example, we see that about 85 percent of the lakes in California subregion and about 40% of the lakes in the Southern Rockies subregion have ANC values below this critical level.

The Direct/Delayed Response Project (DDRP) (Church 1989) was implemented after NSWS sampling and was intended to provide an understanding of the underlying mechanisms responsible for the observed patterns in surface water chemistry. Under DDRP, regional soil surveys were made on statistically representative sets of watersheds consistent with the NSWS, and three models of surface water acidification or recovery (MAGIC, ILWAS, and Enhanced Trickle-Down) were applied in order to predict changes in surface water chemistry likely to occur under various sulfate deposition scenarios (Jenne et al. 1989). In principle, the DDRP could set the primary context for assessment activities related to determining whether patterns and trends in surface water chemistry observed in the TIME network following emissions reductions correspond with DDRP model forecasts for surface water effects. In reality, this task is complicated by the fact that, in order for these comparisons to be valid, emission reductions scenarios would have to exactly match the assumed scenarios actually used in DDRP, which is unlikely to be the case.

The Long Term Monitoring (LTM) project was initiated in 1982 to follow trends in surface water chemistry of systems with low ANC across national and regional gradients of atmospheric sulfate deposition (Newell et al. 1987). This is a cooperative project, in that EPA has supported federal, state, and university researchers in chemical monitoring of 90 lakes and streams. These surface water systems are located across the northeastern and northcentral United States as far west as Colorado (Figure 9.17), and are typically sampled monthly or seasonally. More complete descriptions of the LTM project may be found in Newell et al. (1987).

9.2.3 Conceptual Model of the TIME Project

A working conceptual model for the TIME project used during the 1987–1989 planning period is depicted in Figure 9.18. The left side of the figure represents the existing NSWS program elements upon which the TIME project has been built; the axle represents the temporally intensive LTM project and the wheel is a collective representation of the statistically representative regional surveys. The right side of the figure extends this analogy to the TIME project.

The wheels are designed to provide periodic assessments of status and trends for particular geographic regions and surface water subpopulations of interest. Axle systems will be chosen to represent particular subpopulations of interest, and will provide additional information on seasonal and annual variability necessary to interpret information derived from the regional assessments. For axle systems that are not elements

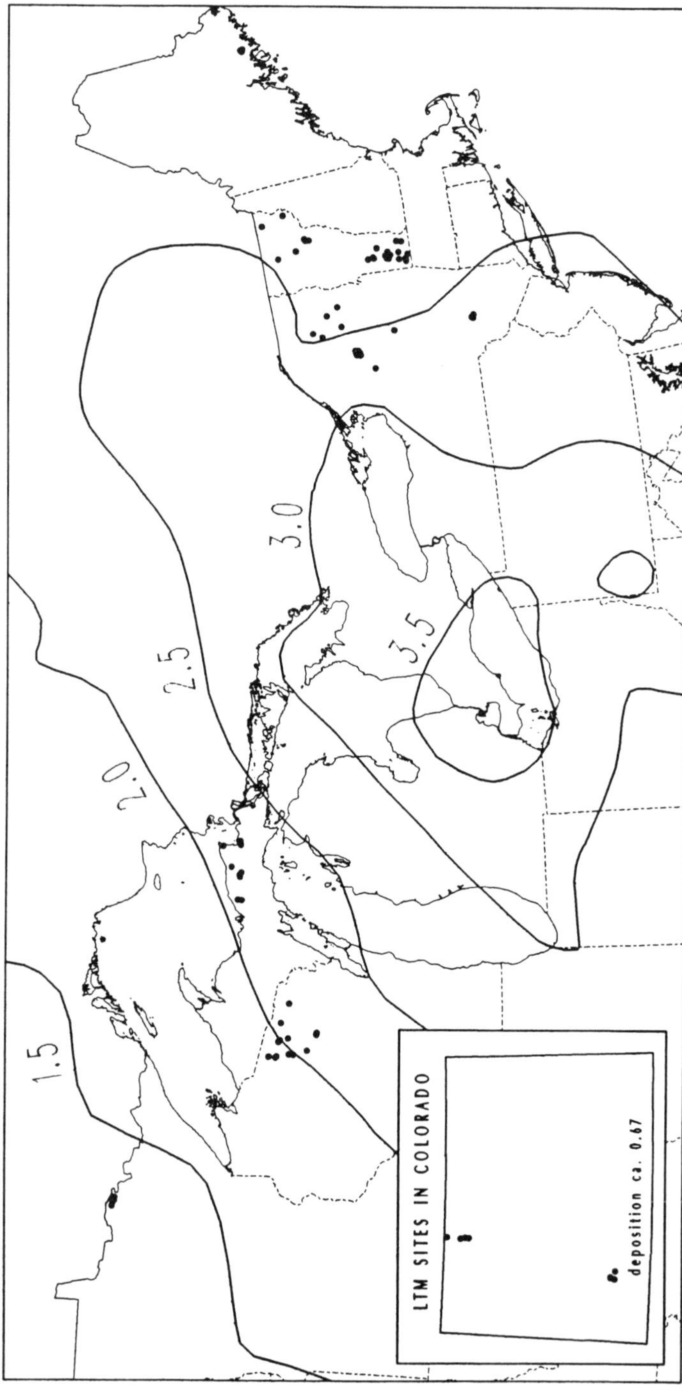

Figure 9.17 Long Term Monitoring project sites with wet sulfate deposition contours (g/m^2 per year, 1980–1984 average) (from Ford et al. 1989).

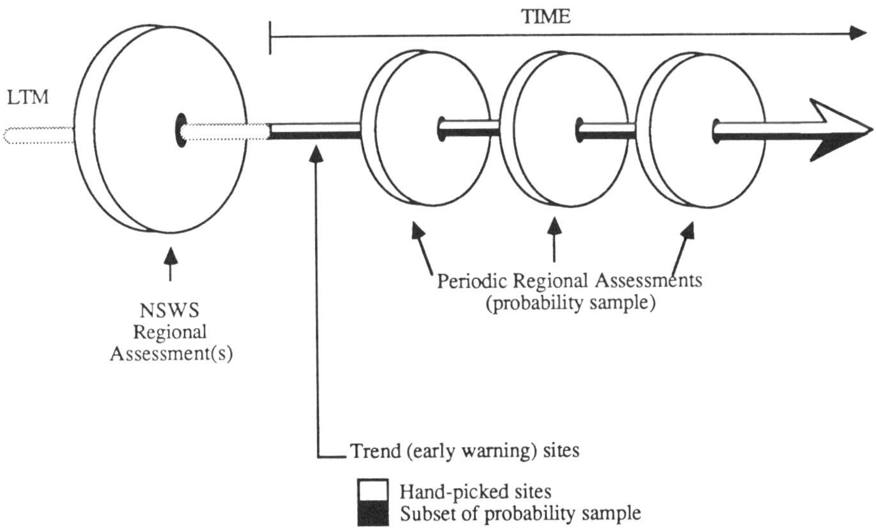

Figure 9.18 Temporally Integrated Monitoring of Ecosystems (TIME) project conceptual "wheel and axle" model.

of the NSWS (referred to as "found" sites), the relationship between these sites and sites in the NSWS population will be described using appropriate statistical approaches (e.g., Overton, 1990a; Ford et al. 1990).

Combining temporally intensive and regionally extensive elements in the TIME design balances the need for regional assessments with the need for ongoing monitoring of rapid response systems for networks addressing specific pollutant impacts such as acidic deposition.

Periodic regional assessments can occur on any sampling schedule, and can easily be developed within the EMAP context as either resurveys of subsets of NSWS sites or resurveys of EMAP sites at grid densities appropriate to the TIME objectives. One possible sampling scheme for the eastern U.S., for example, would be to focus sampling on the DDRP sites in the Northwest, Mid-Atlantic, and Southern Blue Ridge Province, since the DDRP sites are themselves a subset of NSWS sites. Regional assessments can follow the original NSWS protocol, with a single sampling effort during an "index" period of relatively stable water chemistry (e.g., fall turnover for lakes, spring baseflow for streams). Other possibilities for articulating complementary temporally intensive and regionally extensive elements of project design will undoubtedly arise as the EMAP planning and design process continues.

9.2.4 The TIME Data Analysis Plan

The outline for TIMEs Data Analysis Plan (DAP) began in late 1987, and the process to date in working through the elements of the DAP has been valuable in identifying issues that guide monitoring network design. The activities identified in TIME's DAP are currently expected to contribute to the overall Data Analysis Plan for EMAP-Surface Waters. In this section, we discuss the original elements of TIMEs DAP to elucidate the types of planning activities that have thus far influenced the design of the TIME water quality information system.

The purpose of the Data Analysis Plan is to describe the statistical analyses that will be performed to address the major monitoring questions and policy issues of interest. By initiating the DAP activities well in advance of actual field implementation, the information needs of prospective methods of data analysis can be incorporated into the actual design of the monitoring network. The kinds of decisions that could be influenced by these activities include the number and geographic distribution of surface water and deposition sites, the frequency and amount of replication of sampling efforts, and the number and types of quality control samples.

The TIME DAP technical chapters and the primary issues identified for each, are as follows:

- *Trends in individual surface water systems.* How do we assess whether or not there are meaningful statistically significant changes over time in a specific response variable for an individual system? In a vector of variables? Are there advantages to using multivariate, rather than univariate, trend detection techniques?
- *Trends in regional populations.* How do we characterize given response variables for populations of interest in a region or subregion? How do we compare the distribution of a response variable between two time periods? How can univariate and multivariate trends in regional populations of surface water systems be estimated, and how do we assess whether observed trends in response variables are both statistically significant and ecologically meaningful?
- *Trends in biological parameters.* What kinds of information can biological response variables add to a chemical monitoring program? What biological response variables should be used to assess aquatic effects of acidic deposition? Which response variables are best for early-warning signals, which are best for regional assessment of change from one time period to another, and which are best for

detecting biological trends over time? How do we assess, at any given point in time, whether or not these biological response variables have gone below or exceeded a critical level? What statistical techniques are appropriate for the assessment of changes over time in the biological response variables of interest?

- *Integration of non-NSWS surface water systems into NSWS populations.* How do we determine the subsets in the probability sample that are represented by the individual (axle) surface water systems? To what extent can we extrapolate observed changes in individual surface water systems to changes in subpopulations in given regions or subregions? To what extent can we use observed variability in trends in individual systems or groups of axle systems to estimate variability in the subpopulation trends?
- *Trends in deposition.* Is it necessary to co-locate deposition and surface water monitoring sites? Can deposition to axle systems be adequately represented by data from existing deposition monitoring sites? How representative are the regional deposition networks of the TIME regional populations? Is it necessary to estimate and test the statistical and practical significance of trends in deposition at individual stations, and if so what are the best techniques to do so? How do we estimate and test the statistical and practical significance of regional deposition trends?
- *Relationship between trends in deposition and trends in surface water chemistry.* How do we assess whether or not there are statistically and practically significant relationships between changes in deposition and changes in surface water chemistry in an individual surface water system? Between changes in deposition and changes in regional surface water chemistry?

The technical chapters in the Data Analysis Plan are intended to describe and provide sample applications of the approaches recommended for data analysis by cooperating researchers, as well as descriptions of other approaches considered and justifications for the selection of specifically recommended approaches. The assumptions and the sensitivity of analytical results to these assumptions also will be described.

The Data Analysis Plan is intended to be read by a wide audience; thus the recommended techniques will be explained at a level understandable by scientists and decision makers with a knowledge of basic statistics. Appendices to the document will describe the recommended techniques in greater technical detail for those who are interested. In particular, the Appendices are intended to be sufficiently detailed so that data analysts, database managers, and programmers can use them as guides to implementation of recommended techniques.

More detailed descriptions of TIMEs design and discussion of the scientific issues underlying the issues targeted in the Data Analysis Plan can be found in several recent publications. Project history and design are discussed in Ford (1990). Young and Ford (1989) describe a process by which the NSWS data base can be analyzed to optimize site selection for both the early-warning and the regional assessment components of the TIME design. Young (1990) gives a detailed description of a chemical model for evaluating potential sites of interest using this process, and Ford et al. (1990) give applications of this model to data from LTM sites. Overton (1990a) describes a means of relating non-NSWS sites to the NSWS probability sample.

Selections of appropriate statistical techniques for univariate trend detection were studied using a large scale Monte Carlo simulation experiment (Taylor and Loftis 1989). The implications of these results for the TIME project are discussed by Loftis and Taylor (1989). These analyses are currently being extended to consider multivariate trend detection techniques. A related issue explored in connection with these analyses was determining the time to detection of statistically significant trends (given realistic trend magnitudes) at given levels of Types I and II errors, given various combinations of site density and monitoring frequency (Loftis and Taylor 1989). The critical issue in these analyses related to the likelihood of obtaining usable trend information in policy-relevant time frames. Good variance estimates for the chemical response variables of interest were necessary for these kinds of analyses, as were estimates of likely trend magnitudes. In the case of the TIME Project, the former type of information was in fact available by virtue of the combined existence of information from both the NSWS and the LTM Project. The latter type of information, likely trend magnitudes, was bounded by examining DDRP predictions as well as records from a Sudbury lake current experiencing rapid chemical amelioration due to large local decreases in deposition from the Sudbury plants.

Identifying statistical techniques for regional trend detection is a complex task. Preliminary progress was made in this arena in the analysis of the ELS Phase II data (Overton, 1990*b*), in which regional data from four different time periods was available for analysis (spring, summer and autumn of 1986 and, from Phase I, autumn of 1984). Continued development of this area is underway within the context of EMAP.

Hughes (1989) gives an overview of the rationale for biological monitoring, and the usefulness of various specific biological response variables for the particular purposes of the TIME project was considered at the U.S.-Canadian Workshop on Biological Monitoring for Acidification Effects, summarized by Marmorek et al. (1988). Publication of this latter document together with revisions of the eight background papers prepared for the workshop by North American experts in five different organismal groups is in progress [phytoplankton (diatoms, chrysophytes, green and blue-green algae, and dinoflagellates), periphyton, zooplankton, benthic invertebrates, and fish].

Deposition has received relatively less attention, although recently Daly (1989) has outlined a preliminary conceptual approach for estimating wet deposition to TIME regions and individual monitoring sites, and for assessing the adequacy of existing deposition monitoring networks for representing deposition to regions of concern for TIME. Further elaboration of the deposition network design will be undertaken in the context of EMAP-Air/Deposition.

9.2.5 Summary

The design of a national monitoring network to assess the effects of changes in acidic deposition on surface waters of the U.S. is a complex undertaking. EPAs approach has been to embed this design in the context of ten years of NAPAP-related research. Outlining the elements of a Data Analysis Plan for this network helped in the early identification of issues that affect network design, and have allowed technical studies to be undertaken to address pertinent issues. The authors briefly discuss several kinds of analyses that will be used to guide network design decisions of issues such as geographic coverage, site density and selection, monitoring frequency, appropriate response variables, frequency and amount of replication in sampling efforts, and numbers and types of quality control samples.

ACKNOWLEDGMENTS

Reviews by T.C. Young, D.J. Blick, C. Daly, G. Larson, P. Larsen, and S. Paulsen have helped clarify the discussion of ideas in this manuscript. The information in this document has been funded wholly by the U.S. Environmental Protection Agency under contract DW12931230 to The U.S.D.A. Forest Service and 68–03–3439 to Kilkelly Environmental Associates. It has been subjected to Agency review and approved for publication.

REFERENCES

Barchet, W.R. 1987. Acidic deposition and its gaseous precursors. *in* Chapter 5, NAPAP Interim Assessment: The Causes and Effects of Acidic Deposition. Vol. III: Atmospheric Processes and Deposition. U.S. Government Printing Office, Washington, D.C.

Blick, J., J.J. Messer, D.H. Landers, and W.S. Overton. 1987. Statistical basis for the design and interpretation of the National Surface Water Survey, Phase I: lakes and streams. *Lake Reserv. Manage.* 3:470–475.

Church, M.R. 1989. Direct/Delayed Response Project: Predicting future long-term effects of acidic deposition on surface water chemistry. *EOS* 70:801–813.

Daly, C. 1989. Estimation of wet deposition to surface water systems for the TIME Project: A geographic approach. pp. 436–446 *in* R.C. Ward, J.C. Loftis, and G.B. McBride (editors). *Proceedings International Symposium on the Design of Water Quality Information Systems.* Information Series No. 61. Colorado Water Resources Research Institute. Colorado State University, Fort Collins, CO.

Driscoll, C.T., G.E. Likens, L.O. Hedin, J.S. Eaton, and F.H. Bormann. 1989. Changes in the chemistry of surface waters. *Environ. Sci. Technol.* 23(2):137–143.

Eilers, J.M., D.F. Brakke, D.H. Landers, and P.E. Kellar. 1988. Characteristics of lakes in mountainous areas of the western United States. *Verh. Internat. Verein. Limnol.* 23:144–151.

EPA 1990. Environmental Monitoring and Assessment Program. Overview. EPA/600/9–90/001. U.S. Environmental Protection Agency Office of Modeling, Monitoring Systems, and Quality Assurance (RD-680). Washington, D.C.

Ford, J., T. Young, and J. Stoddard. 1990. Long-term monitoring for aquatic effects of acidic deposition: An approach to regionalizing information from existing monitoring sites. *Proceedings of the Workshop on Assessment and Monitoring of Acidification of Rivers and Lakes,* Freiburg, FRG. 17–19 October 1989.

Ford, J., D. Landers, and J. Eilers. 1988. "Kenai Lakes Investigation Project (KLIP): Acid Rain Research in Alaska." Ann. Mtng. Am. Assn. Adv. Sci.–Arctic Div. Fairbanks, AK, October 1988.

Hedin, L.O., G.E. Likens, and F.H. Bormann. 1987. Decrease in precipitation acidity resulting from decreased SO_4 concentration. *Nature* 325:244–246.

Hughes, R.M. 1989. What can biological monitoring tell us about environmental health? pp. 447–460 *in* R.C. Ward, J.C. Loftis, and G.B. McBride (editors). *Proceedings International Symposium on the Design of Water Quality Information Systems.* Information Series No. 61. Colorado Water Resources Research Institute. Colorado State University, Fort Collins, CO.

Husar, R.B. 1986. Emissions of sulfur dioxide and nitrogen oxides and trends for eastern North America. In: *Acid Deposition: Long-Term Trends* (pp. 48–92). Committee on Monitoring and Assessment of Trends in Acid Deposition. National Academy Press, Washington, D.C.

Jenne, E.A., L.E. Eary, L.W. Vail, D.C. Girvin, A.M. Liebtrau, L.F. Hibler, T.B. Miley, and M.J. Monsour. 1989. "An Evaluation and Analysis of Three Dynamic

Watershed Acidification Codes (MAGIC, ETD, and ILWAS).'' Pacific Northwest Laboratory (PNL-6687).

Kaufmann, P.R., A.T. Herlihy, J.W. Elwood, M.W. Mitch, W.S. Overton, M.J. Sale, J.J. Messer, K.A. Cougan, D.V. Peck, K.H. Reckhow, A.J. Kinney, S.J. Christie, D.D. Brown, C.A. Hagley, and H.I. Jager. 1988. *Chemical Characteristics of Streams in the Mid-Atlantic and Southeastern United States. Volume 1. Population Descriptions and Physico-Chemical Relationships*. EPA/600/3–88/021a. U.S. Environmental Protection Agency, Washington, DC.

Landers, D.H., J.M. Eilers, D.F. Brakke, W.S. Overton, P.E. Kellar, M.E. Silverstein, R.D. Schonbrod, R.E. Crowe, R.A. Linthurst, J.M. Omernick, S.A. Teague, and E.P. Meier. 1987. *Characteristics of Lakes in the Western United States. Volume 1. Population Descriptions and Physico-Chemical Relationships*. EPA/600/3–86/054a. U.S. Environmental Protection Agency, Washington, DC.

Landers, D.H., W.S. Overton, R.A. Linthurst, D.F. Brakke. 1988. Eastern Lake Survey. *Environ. Sci. Technol.* 22(2):128–135.

Landers, D.H. et al. 1989. ''Regional Evaluation of Lakes on the Kenai Peninsula of Alaska.'' 42nd Annual meeting, American Society of Limnology and Oceanography, Fairbanks, AK, June 1989.

Linthurst, R.A., D.H. Landers, J.M. Eilers, D.F. Brakke, W.S. Overton, E.P. Meier, and R.E. Crowe. 1986. *Characteristics of Lakes in the Eastern United States. Volume 1. Population Descriptions and Physico-Chemical Relationships*. EPA/600/4–86/007a. U.S. Environmental Protection Agency, Washington, DC.

Loftis, J.C. and C.H. Taylor. 1989. Detecting acid precipitation impacts on lake water quality. Environmental Management, Vol 13(5):529–538.

Loftis, J.C., R.C. Ward, R.D. Phillips, and C.H. Taylor. 1989. An evaluation of trend detection techniques for use in water quality monitoring programs. EPA/600/3–89/037. Environmental Protection Agency, Washington, D.C.

Marmorek, D.R., D.P. Bernard, and J. Ford. 1988. Biological monitoring for acidification effects: results of a U.S.-Canadian workshop. Internal report.

Messer, J.J., C.W. Ariss, J.R. Baker, S.K. Drouse, K.N. Eshleman, P.R. Kaufmann, R.A. Linthurst, J.M. Omernick, W.S. Overton, M.J. Sale, R.D. Schonbrod, S.M. Stambaugh, and J.R. Tuschall, Jr. 1986. *National Stream Survey Phase I - Pilot Survey*. EPA/600/4–86/026. U.S. Environmental Protection Agency, Washington, DC.

Messer, J.J., C.W. Ariss, J.R. Baker, S.K. Drouse, K.N. Eshleman, A.J. Kinney, W.S. Overton, M.J. Sale, and R.D. Schonbrod. 1988. Stream chemistry in the Southern Blue Ridge: feasibility of a regional synoptic sampling approach. Water Resources Bulletin 24(4):821–829.

Newell, A.D., C.F. Powers, and S.J. Christie. 1987. *Analysis of Data from Long-Term Monitoring of Lakes*. EPA/600/4–87/014. U.S. Environmental Protection Agency, Washington, DC.

Pollack, A.K. and J. Ford. 1990. The TIME Project: An Overview. pp. 413–424 *in* R.C. Ward, J.C. Loftis, and G.B. McBride (editors). Proceedings International Symposium on the Design of Water Quality Information Systems. Information Series No. 61. Colorado Water Resources Research Institute. Colorado State University, Fort Collins, CO.

Omernik, J.M., and C.F. Powers. 1983. Total alkalinity of surface waters—a national map. Annals of the Association of American Geographers 73:133–136.

Omernik, J.M., and A.J. Kinney. 1985. Total Alkalinity of Surface Waters: A Map of the New England and New York Region. EPA/600/D-84/216. U.S. Environmental Protection Agency, Corvallis, OR.

Omernik, J.M., and G.E. Griffith. 1986a. Total alkalinity of surface waters: a map of the Upper Midwest Region of the United States. Environmental Management 10:829–839.

Omernik, J.M., and G.E. Griffith. 1986b. Total Alkalinity of Surface Waters—A Map of the Western Region. EPA/600/D-85/219. U.S. Environmental Protection Agency, Corvallis, OR.

Overton, W.S. 1990a. "A strategy for the use of found sites in a rigorous monitoring program." Technical Report No. 139, Statistics Department, Oregon State University, Corvallis, OR.

Overton, W.S. 1990b. "Calibration methodology for the double sample structure of the National Lake Survey Phase II sample." Technical Report No. 130, Statistics Department, Oregon State University, Corvallis, OR.

Payne, F.E. and J. Ford. 1988. "The Concept of TIME—Temporally Integrated Monitoring of Ecosystems. Supplement." Internal report. U.S. Environmental Protection Agency Environmental Research Laboratory, Corvallis, OR.

Sanders, T.G., R.C. Ward, J.C. Loftis, T.D. Steele, D.D. Adrian, and V. Yevjevich. 1983. Design of Networks for Monitoring Water Quality. Water Resources Publications. Littleton, CO 328 pp.

Schindler, D.W. 1988. Effects of acid rain on freshwater systems. *Science* 239:149–157.

Taylor, C.H. and J.C. Loftis. 1989. Testing for trend in lake and ground water quality time series. Water Resources Bulletin 25:715–726.

Thornton, K.W., F.E. Payne, J. Ford, and D.H. Landers. 1987. "The Concept of TIME." Internal report. U.S. Environmental Protection Agency Environmental Research Laboratory, Corvallis, OR.

Young, T.C. and Ford, J. 1989. Site selection procedures for Temporally Integrated Monitoring of Ecosystems (TIME). *in* R.C. Ward, J.C. Loftis, and G.B. McBride (editors). *Proceedings International Symposium on the Design of Water Quality Information Systems.* Information Series No. 61. Colorado Water Resources Research Institute. Colorado State University, Fort Collins, CO.

9.3 INITIATION OF A NATIONAL WATER QUALITY NETWORK FOR NEW ZEALAND

(Prepared by D.G. Smith and G.B. McBride, Water Quality Centre, Department of Scientific and Industrial Research, Hamilton, New Zealand)

The two main islands of New Zealand (34°S—47°S; 166°E—179°E, surface area = 265,000 km^2, population about 3.2 million) have a mountainous elongated character and lie in the path of wet westerly winds from the Tasman Sea. Average rainfall and runoff is about 1500 mm and 47.5 L/s/km^2, but these are distributed rather unevenly, being larger on the west coasts. Rivers are typically steeper than those in continental countries. There are a number of large inland lakes. The economy is still dominated by agricultural production: sheep, dairy and cattle farming, and some horticulture. These operations have caused the majority of water quality problems, both from point and diffuse sources: dissolved oxygen deficits, sewage fungus patches, and lake eutrophication. Problems associated with industry tend to be located in waters near the main cities. Most of these cities are near estuaries and coasts.

In mid-1988 the Water Quality Centre received funding to design a network for rivers and streams, building on a previous draft design for rivers and lakes (McBride 1987). This task has been completed and the network began operation in 1989. The design included lakes, but this is yet to be implemented.

The following lists the tasks which have been considered in the design, all of which will receive attention here. Full details are to be found in Smith et al. (1989).

Tasks to be addressed

1. Develop the goal and objectives
2. Produce a list of sampling sites
3. Confirm statistical design criteria, especially sampling frequency
4. Produce a list of determinands
5. Assess accuracy requirements
6. Confirm laboratory analysis procedures
7. Recommend appropriate laboratories
8. Recommend field operators and training requirements
9. Develop field sampling/analysis procedures
10. Develop procedures for sample preservation/transport to laboratory
11. Develop sampling routes
12. Develop quality assurance procedures (from bottle washing to data entry
13. Recommend data storage/retrieval/management system
14. Recommend data analysis procedures and software
15. Recommend reporting procedures (what, to whom, by whom, when)
16. Assess cost

9.3.1 Goals and Objectives

The *goal* of the network is to provide scientifically defensible information on the important physical, chemical, and biological characteristics of a selection of the nation's river waters as a basis for advising the Minister of Science and other ministers of the crown of the trends and status of these waters.

The *objectives* are to establish a national water quality database such that one can:

1. Detect significant trends in water quality, and
2. Develop better understandings of the nature of the water resources and hence to better assist their management.

With respect to the first objective, the intention is to be able to detect a monotonic trend equal to the standard deviation of the detrended data over a five year period, if such a trend occurs.

9.3.2 Sampling Sites

There are two types of stations in the network: (1) "Baseline" (where there is likely to be no or little effect of diffuse or point source pollution and which will account for natural or near-natural effects and trends); and (2) "Impact" (which are downstream of present, and possible future, areas of agriculture, afforestation, industry and urbanization). Both types are required to properly address trends but in some instances only baseline or impact stations are possible. The requirements for station selection are that they should be:

1. At or close to a current hydrological recording station.
2. Accessible.
3. At such a location so that transport of samples to the laboratory presents no time constraints.
4. Safe for operators.
5. Free from disturbing influences such as weirs and weedbeds.
6. Easy to sample and carry out *in situ* work.
7. Free from saline intrusions.
8. In some way of national, multiregional or scientific importance.

Most sampling sites have been visited by the designers. Sites have been selected so that a single surface grab sample will be representative of the bulk water flowing past. In all, there are 77 river sites, 44 in the North Island and 33 in the South Island (Figure 9.19—which shows lake sites included in the design). The inclusion of each site has been clearly justified.

Figure 9.19 Sampling sites. River sites are marked with a bar and coded, lakes are named.

9.3.3 Statistical Design Criteria

The design requirements, which include the analysis of data, are discussed in a later section. The authors briefly examine the statistical reasoning behind the choice of sampling frequency for the most important objective, that of trend detection.

When trends in data are sought, it must be recognized that there is

Figure 9.19 *Continued*

no point in carrying out short-term intensive sampling, because the effects of seasonality, extreme events, non-uniform variance, and autocorrelation must be accounted for (Lettenmaier 1976, 1978; Loftis and Ward 1980). The practical consequence is (Lettemaier et al. 1982, pp. 62–63) that it is generally difficult to detect a trend on the order of the water quality variable's standard deviation for *n* smaller than 50–100. Also, because of autocorrelation, the minimum effective sampling interval is on the order of two weeks (for rivers monthly sampling is often optimum; for lakes, even longer). So for a trend to be detected, the network must stay fixed for at least five years. This finding is strengthened by Hirsch and Slack (1984), of the United States Geological Survey (USGS), in their examination of a robust nonparametric trend test—reasonable power for trend detection for rivers may only be attainable after five years of sampling. Indeed it is only after ten or so years of routine river data collection that the USGS is publishing results of trend assessments (Smith and Alexander 1983, Smith et al. 1987). Lettenmaier et al. (1982) go so far as to conclude that revisions of the network should *not* take place annually, because this encourages fragmentation of the data, which compromises its utility for trend detection. Analysis of results of the network should be performed annually, however, to look for any evidence of trends emerging. Provision of such analysis software within the proposed data management system (see later) is thus desirable.

The technique here is basically that promoted by Lettenmaier (1976, 1978). The simple null hypothesis is proposed:

$$H_0\text{: } \Delta\mu = 0 \qquad (9\text{-}1)$$

where $\Delta\mu$ is the change in the mean (μ) of a water quality variable, against the composite alternative hypothesis

$$H_A\text{: } \Delta\mu \neq 0 \qquad (9\text{-}2)$$

Because H_A is composite, the power of a test to choose between (9–1) and (9–2) is a function of $\Delta\mu$. We consider $\Delta\mu$ to be monotonic–either a step or a linear trend (higher order monotonic polynomial trends are also possible in this technique).

Our method of approach was to calculate the detection threshold ($\Delta\mu$/s, where s is the detrended standard deviation) of a test for given Type I and Type II error risks, trend type, autocorrelation, sampling frequency and length of record. Thresholds have been calculated for a given combination of α (the probability of committing the Type I error of falsely detecting a monotonic trend), β (the probability of committing the Type II error of failing to detect a monotonic trend), r_1 (the daily lag 1 autocorrelation coefficient), and sampling frequencies over 5 and 10 year periods. The authors also examined autocorrelation coefficients for a variety of determinands measured during long term studies (McBride 1987).

Our conclusion from this is that an appropriate sampling frequency for linear trend detection is monthly for rivers and bimonthly for lakes to obtain the desired detection threshold (of around unity) within a 5 year period with reasonable power ($\beta \cong 0.1$).

9.3.4 Variables to Be Measured (Determinands)

All determinands have been carefully selected and included only if there are very good reasons for doing so. The temptation to include "nice-to-do" determinands has been avoided. Table 9.1 contains the selected determinands, where they are to be measured, whether they are for rivers or lakes, and the reason for inclusion of each.

9.3.5 Accuracy Requirements

The network has been designed to give good trend detectability for regular sampling over 5–10 years. The magnitude of the trend to be detected, either gradual or abrupt, is measured by the difference in expected values of observations near the start of the trend assessment period and those near the end of that period. We want this minimum detectable trend to be as small as possible, minimizing the impact of analytical and sampling errors. We cannot, however, design for the magnitude of the trend on its own—we have to design for the minimum detectable trend as a proportion of the standard deviation of the detrended data. And this standard deviation contains a component attributable to the unavoidable presence of errors, both analytical and sampling. So the larger the error standard deviation, the larger is the minimum detectable trend. Because we want this trend to be as small as possible, a way has been found to: (a) calculate the effect of errors on the minimum detectable trend, and (b) indicate to the analysts and samplers what sort of performance is required.

The details of how this is done are given elsewhere (Smith et al. 1989). The essential idea is to compare an attainable error standard deviation with a maximum permissible error standard deviation. If the

Table 9.1. List of variables to be measured.

DETERMINAND	WHERE MEASURED (FIELD/BASE/LABORATORY)	RIVERS/ LAKES	REASON FOR INCLUSION
Dissolved oxygen	f	R/L	Numerical standard; necessary requirement for aquatic life; rapid indicator of organic pollution; indicator of lake trophic state.
pH	f /b /l*	R/L	Numerical standard; aquatic life protection: pollution indicator: acidification.
Conductivity	l	R/L	Simple surrogate for total inorganic ions (incl. hardness).
Temperature	f	R/L	Numerical standard; DO interpretation; mixing in lakes: aquatic life protection.
Visual clarity (black disc); Secchi disc depth	f f	R (initially†) L	Descriptive standard: visual monitor; erosion/habitat destruction; aquatic life protection (clarity only); public perception.
Turbidity and absorption coefficients	l	R/L	Descriptive standard. Turbidity as support to on-site clarity measurement, absorption coefficients relate to water colour, light climate for plants and organic character of water.
BOD_5	l	R	Descriptive standard; organic pollution indicator.
Dissolved nutrients: NH_3/NO_3/DRP	l	R	NH_3 is proposed numerical standard for potable supply: NO_3 required for public health issues; enrichment potential for algal growth.
TP/TN	l	R/L	Nutrient status/ eutrophication.
Flow	f	R	Flow dependence analysis.
Lake height	f	L	Height dependence analysis.
Chlorophyll *a*	l	L	Descriptive standard: possible water use change indicator; nutrient status indicator.

Table 9.1. *Continued*

DETERMINAND	WHERE MEASURED (FIELD/BASE/LABORATORY)	RIVERS/ LAKES	REASON FOR INCLUSION
Benthic invertebrates**	l	R	Descriptive standard; water use change indicator.
Nuisance periphyton growths	f	R	Descriptive standard; visual monitor; public perception
The following for first year only:			
Ca/Mg/K/Na/Cl	l	R /L	Major ions including hardness species; general assessment of water quality; HCO_3 drop and SO_4 rise indicators of lake acidification.
HCO_3 /SO_4	l	R /L	

The term 'standards' refers to the current and proposed standards.
The 'first year only' ions should be measured at a later stage only if conductivity appears to be changing. There is one exception to this; for lakes it may be wise to measure HCO_3 and SO_4 in each sample as a reassurance that lakes are not being acidified.
No bacterial measurements are recommended at present. Faecal coliforms are stipulated in legislation but the wisdom of this is being seriously questioned. For instance EPA (1986) recommend *E. coli* and enterococci for freshwater bathing criteria. In addition there are problems with sampling frequency differences between legislation and what is practical for a national network. There are also difficulties with interpretation. We will await results of research before embarking on a costly analysis which is of dubious value.
* pH has been measured in the field, back at survey base, and in the laboratory (as an additional check) for the first year. The 3 sets of measurements were examined and as a consequence pH is now measured only in the laboratory.
** Benthic invertebrates can be collected at about 70 of the proposed river sites and data will be analysed for changes in community composition and species abundance.
† Black disc measurements will be made initially only on rivers. When current research is complete, a recommended method for horizontal black disc distance in lakes will be available. Both vertical and horizontal measurements will then be made.

former error is the smaller, then the trend can be detected with the required power. The analysis is based on the assumption that we can ignore the effect of bias (because trend detection involves differences of means, and so most if not all bias will cancel out). Hence the analysis considers only precision and considers laboratory precisions based on many years of analytical quality assurance studies and anticipated field precisions based on best judgement where actual values are not available.

The upshot of this part of the design is that we are in an *a priori* position to state the actual trend detectability for a variety of determinands and certain concentrations (or values), given certain values for α and β and allowing certain permissible increases in detectable trend caused by errors.

To ensure that the laboratory is producing accurate results, an appropriate quality control (QC) program is required which includes the usual within-laboratory protocols and comparative tests with other laboratories. It is also necessary to ensure that field analysis produces accurate results and that field sampling does not introduce inaccuracy. A comprehensive data quality assurance program is called for and has been built into the design.

9.3.6 Laboratory Analysis Procedures

Sample processing flow sheets have been drawn up and a list of preferred analytical methods established. The flow sheet for riverine samples is presented in Figure 9.20.

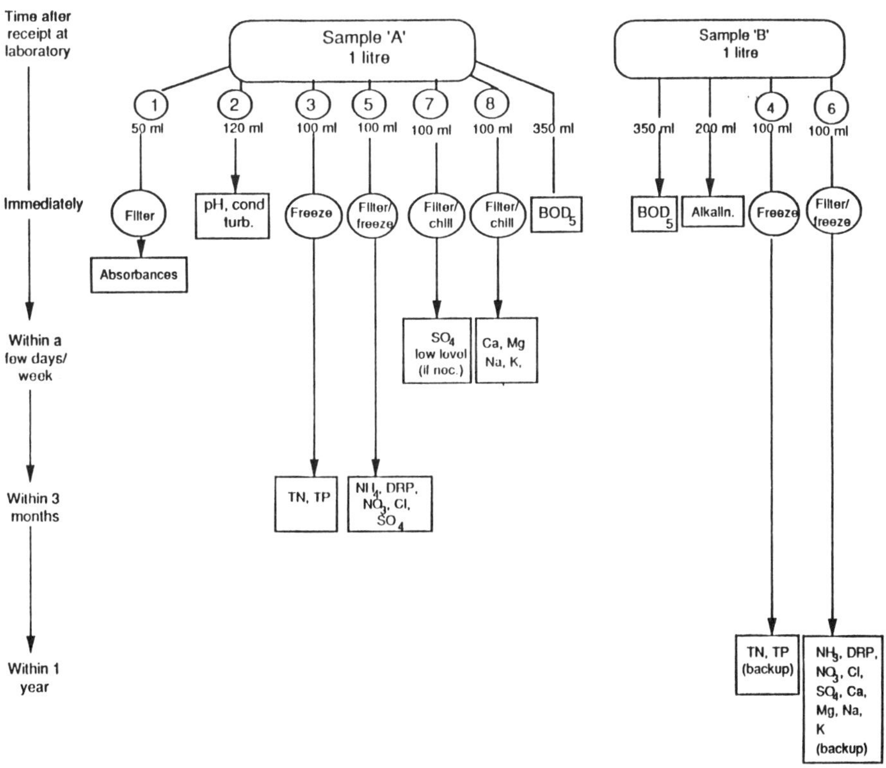

Figure 9.20 Laboratory sample processing flow sheet for rivers.

9.3.7 Laboratory and Field Operator Training Requirements

Several studies have shown the difficulties associated with the use of a multitude of different agencies for a national water quality network. For instance in the UK, Fenlon and Young (1982) noted that the harmonization of results (i.e., the production of results of known and adequate accuracy), which is difficult on the regional scale, has proved to be even more so at the national level and had been achieved to only a limited degree *seven* years after the inception of the UK Harmonized Monitoring scheme. In addition, there is also no harmonization in the manner in which data are supplied to the central data repository (the Department of the Environment) in large part because there are 16 autonomous contributing agencies.

In the USA, the General Accounting Office (1981) has stated that "inconsistency in field work and laboratory performance make it virtually impossible to meaningfully compare network data from month to month, season to season, and year to year."

Consequently, a single agency (the DSIRs Water Resources Survey field party network) has been assigned the task of field work (i.e., water sampling and field analysis—which has been kept to a bare minimum), and just one laboratory has been selected for the laboratory analysis. It is possible that other laboratories could join the network later, if the necessary standards are attained.

The training requirements of field staff have been addressed by holding a refresher training course to supplement previous courses and video training films. At the course emphasis was placed on correct field procedures for ensuring accurate and precise measurements for dissolved oxygen concentration, pH, and temperature; contamination-free water sampling and subsampling; and benthic invertebrate sampling. As a consequence of the course, modifications have been built into the recommended methods. For example, calibrated thermometers to check field probes (all too easily overlooked) are used and known pH low concentration sodium bicarbonate solutions (equilibrated with atmospheric CO_2) to simulate natural waters as a check on the functioning of pH probes are used. This is in *addition* to the normal calibration procedure which by itself is inadequate when dealing with poorly buffered, low ionic strength natural water samples. Without such checks little reliance can be placed on the reliability of results from the apparently simple task of making temperature and pH measurements.

As noted in Table 9.1, pH measurements are no longer made by field staff but the same low ionic strength known pH solutions are used by the laboratory. We have found minimal (often < 0.1 unit) pH differences between field and next-day laboratory measurement. This, together with the major time demand required by field staff for accurate field pH measurement (a very difficult task for natural waters and one which is often done without a full appreciation of the difficulty), resulted in the decision to curtail such measurements.

9.3.8 Sampling Routes and Sample Preservation/Transport

Once the sampling sites had been established, each of the 15 field parties was asked to consider an appropriate sampling route to ensure that each river site was visited at approximately the same time of day on each visit and that site visitation was keyed into their already established work load (i.e., their hydrological station visits/maintenance). At this stage it was stressed that it was probably better to avoid doing the water quality work as an add-on to their normal work load unless it was certain that the additional effort was not likely to compromise the water quality work.

It is important that samples can be transported overnight to the laboratory because sample preservation is achieved simply by chilling under ice in an insulated box. An overnight commercial air courier service has been built into each sampling route to ensure this occurs.

9.3.9 Field Sampling and Analysis Procedures

The location of each sampling site has been specified. The requirements for each site have been defined for field parties and reinforced by carefully thought out field sheets (see Figure 9.21 for an example) which must be completed at the time of sampling. The design of the field sheets has been carried out with the cooperation of the field parties—indeed the cooperation of, and input from, field parties has been sought at all appropriate stages in the network design. This was felt to be important.

At each site, each month, representative water samples are collected in each of two one litre containers. Dissolved oxygen and temperature are also determined in the field. Water clarity is determined using a specially constructed black disc viewed horizontally (Davies-Colley, 1988). This provides a more appropriate measure of clarity than the conventional Secchi disc depth because it relates better to the sighting

(This sheet must accompany samples dispatched to the laboratory for analysis - one sheet per site per visit)

Day................................ Date................................ Field party area................

Field operators' names ....Time (NZST):

Site code: Site Description: Cloud (/8):

Gauge ht (m) (external): (internal): Flow (m^3/s):

Water temp (°C) (corrected): DO (% satn): Atmos.press. (mbar)(±5)

Black disc dist (m): (a) pH (field): pH (base):
(2 observers) (b) 8.28 check: 8.28 check:

	Obs.	1	2	3	4	5	6	7	8	9	10
% peri-phyton cover	Filament-ous										
	Mats (>2 mm thick)										

Periphyton colour (filamentous): (mats):
Observation site (circle one): open/shaded
Comments:

Notes:

1 At each site measure and record temperature, pH, DO and horizontal black disc distance. At end of day re-measure pH on samples collected in 120 ml containers. Also, obtain and record values for gauge ht, flow and atmospheric pressure.

2 At each site fill two 1 litre plastic containers and one 120 ml container. Cap tightly, leaving no air space if possible.

3 Assess % periphyton cover at 10 equidistant points across the wadeable area of a 'run' near the water sampling site. Improvise as necessary to get 10 observations.

4 Comments could concern: exceptional low/high flow conditions and antecedent weather; surface scums, foams, oils and slicks; anything unusual, e.g. different or unusual colour.

Figure 9.21 Original field sheet for rivers.

range of riverine animals, gives a measure of the beam attenuation coefficient, and finally it can actually be measured in a shallow river. Finally an assessment is made of nuisance periphyton cover. Each year an appropriate number of samples are taken of benthic invertebrates.

For all the above requirements, detailed instructions have been sent to and discussed with field parties.

9.3.10 Quality Assurance Procedures

Here we are referring to all aspects of ensuring that measurements made in the field and laboratory are within accuracy specifications, and that transposition of data and loading onto computer files occurs without error.

For field measurements, quality assurance procedures have been built into each method, and each sampling bottle has been coded and is dedicated to a particular site.

The laboratory is expected to maintain normal quality assurance procedures; i.e., by using appropriate blanks and known solutions, and production of control charts as appropriate. In addition, test solutions (these can be either complete unknowns or known solutions) are periodically given to the laboratory for analysis. Even the techniques for routine bottle washing have been provided for use by the laboratory.

Built into the data processing package (AQUAL) is a checking system to ensure that all results entered could be reasonably expected for New Zealand's fresh, surface waters. In addition, prior to data entry they are sighted by an experienced staff member who looks for oddities.

9.3.11 Data Management System

There is a dearth of water quality data archive software available. As a result, data documentation and availability are often compromised badly—meaning that the information content of the data is not extracted. It may not even be known that data exist. The authors attempted to avoid this problem in New Zealand by producing a water quality database program—AQUAL (AQuatic Unified Archival Library)—McBride 1989.

This program enables users to store, interpret, retrieve and exchange water quality data. It has been designed to do so in a way that makes using the program as easy and as flexible as possible. At the same time it ensures that sufficient information is given so that data are stored and retrieved unambiguously. The user interface consists of menus of options, full-screen data entry, and prompts. Entry into some screen fields is mandatory—to avoid ambiguity problems. If an input error is made, the user is prompted with a helpful error message. Default entries appear in appropriate fields automatically, and can be overwritten.

9.3.12 Data Analysis

Any data analysis protocol must contain three elements (Ward et al. 1988): (a) procedures to prepare a raw data record for statistical analysis; (b) means to display the behavior of the water quality variables—over time, for trend analysis; (c) recommended statistical analysis procedures to provide the desired information.

In the case of the New Zealand National Water Quality Network, these three elements are fused into two parts:

- (i) regular checks within each year for consistency, errors, and outliers; and
- (ii) yearly reports of the features of the data to date, including emerging trends.

Part (i) includes all of element (a) and some of (b)—the display of the raw data. Part (ii) covers all of element (c) and the rest of element (b)—the display of data interpreted using the (c) procedures.

Note that these yearly reports will become more substantial as time goes by, because trend detection power will increase with the length of record.

Part (i) (regular within-year) checks can be carried out using AQUAL. In particular, the summary statistics and stem-leaf diagrams in AQUAL should be used. These will show up any outliers. If appropriate a plot of the attribute value versus time could be obtained also.

Part (ii) annual reports, excluding trend analysis, can be obtained from the data checking and user-defined report AQUAL options. We expect that heavy use will be made of the summary statistics feature of the user defined report.

For trend analysis, we are recommending following mostly the data analysis procedure used by the USGS (Crawford et al. 1983). This procedure is well written and has been used successfully to report on trends in their NASQAN and NWQSS networks (Smith et al. 1987). Consideration will also be given to using WQSTAT II (Loftis, 1989).

The analysis will be based on the following sequential steps:

1. Compute flow adjusted data (river sites only);
2. Visually inspect the raw and any flow adjusted data for each site;
3. Perform appropriate nonparametric statistical tests, indicated by the visual inspection, on the raw data and on any flow adjusted concentration (FAC) data.

In performing trend tests, nonparametric procedures offer the following advantages (e.g., Lettenmaier et al. 1982, Hirsch et al. 1982, Helsel 1987, Ward et al. 1988):

- they are always valid, whatever the distribution of data happen to be;
- they are nearly as powerful as parametric tests when the assumptions of such tests (e.g., sampling is from normal populations) are fully satisfied;
- they are more powerful than parametric tests when these assumptions are violated, the more so as the violation of the assumptions becomes more extreme;
- they are relatively (sometimes completely) unaffected by outliers, missing data, or non-detects (parametric tests can be affected seriously by such data);
- they are conceptually simple, and computations are relatively straightforward.

It must always be borne in mind that the network cannot be expected to indicate the cause of emerging trends. If an emerging trend is of concern, the reasons for it will have to be investigated by other methods which may involve a quite different sampling program. The network gives an early warning signal for emerging trends.

9.3.13 Reporting

The goal of the network requires reporting to the Minister of Science and other government agents. The first report will comment mainly on the mechanics of the first year's operation. Some summary statistics could be produced but in all probability some of the first year's data may not be ideal and an initial ''settling-down'' period may be required. This first report may not be appropriate for senior government officials.

Thereafter, reports will become more detailed and comprehensive as data accumulate. A recommendation on the continuance (or otherwise) of the network should be made on a periodic basis. Outside parties ought to feel free to request information which should be made available only after it has been carefully prepared. Data released prior to being carefully prepared should be done so with extreme caution and accordingly flagged.

9.3.14 Costs

After the first year, when there are numerous setting up costs (e.g., purchase of field gear), the total cost of the network will be in the vicinity of U.S. $230,000. This is just 0.03% of the combined value of domestic and international tourism (in which the natural water resources play a major role).

REFERENCES

Crawford, C.G., J.R. Slack, and R.M. Hirsch. 1983. Nonparametric tests for trends in water-quality data using the Statistical Analysis System. Open-File Report 83–550. Reston, Virginia: U.S. Geological Survey.

Davies-Colley, R.J. 1988. Measuring water clarity with a black disc. Limnology and Oceanography 33(4):616–623.

Environmental Protection Agency. 1986. Quality criteria for water. Report EPA 440/5–86–001. Washington, D.C.: U.S. Environmental Protection Agency.

Fenlon, J.S., and D.D. Young. 1982. Chemical surveillance of rivers. *Water Pollution Control* 81:343–357.

General Accounting Office. 1981. Better monitoring techniques are needed to assess the quality of rivers and streams (2 vols.). Report CED-81–30. Washington D.C.: US General Accounting Office.

Helsel, D.R. 1987. Advantages of nonparametric procedures for analysis of water quality data. *Hydrological Sciences Journal* 32(2):170–190.

Hirsch, R.M., J.R Slack, and R.A. Smith. 1982. Techniques of trend analysis for monthly water quality data. *Water Resources Research* 18:107–121.

Hirsch, R.M., and J.R. Slack. 1984. A nonparametric trend test for seasonal data with serial dependence. *Water Resources Research* 20:727–732.

Lettenmaier, D.P. 1976. Detection of trends in water quality data from records with dependent observations. *Water Resources Research* 12:1037–1046.

Lettenmaier, D.P. 1978. Design considerations for ambient stream quality monitoring. *Water Resources Bulletin* 14:884–892 (and discussion in 15:1781–1787).

Lettenmaier, D.P., L.L. Conquest, and J.P. Hughes. 1982. "Routine streams and rivers water quality trend monitoring review." Technical Report 75. Seattle: Charles W. Harris Laboratory, University of Washington.

Loftis, J.C., and R.C. Ward. 1980. Water quality monitoring—some practical sampling considerations. *Environmental Management* 4(6):521–526.
Loftis, J.C. 1989. WQSTAT II: a data analysis software package developed for water quality data. Proceedings of the International Symposium on the Design of Water Quality Information Systems, Information Series No. 61, Colorado Water Resources Research Institute, Colorado State University, Fort Collins, Colorado, pages 363–366.
McBride, G.B. 1987. Draft proposal for a national water quality monitoring programme for New Zealand. Internal Report 87/07. Hamilton, New Zealand: Water Quality Centre, Division of Water Sciences, Department of Scientific and Industrial Research.
Smith, D.G., G.B. McBride, G.G. Bryers, R.J. Davies-Colley, J.M. Quinn, and W.M. Vant. 1989. Design for a national water quality monitoring network for New Zealand. Consultancy Report 8016/2. Hamilton, New Zealand: Water Quality Centre, Divison of Water Sciences, Department of Scientific and Industrial Research.
Smith, R.A., and R.B. Alexander. 1983. Evidence for acid-precipitation-induced trends in stream chemistry at Hydrologic Bench-Mark stations. USGS Circular 910. Reston, Virginia: U.S. Geological Survey.
Smith, R.A., R.B. Alexander, and M.G. Wolman. 1987. Water-quality trends in the nation's rivers. *Science* 235:1607–1615.
Ward, R.C., J.C. Loftis, H.P. DeLong, and H.F. Bell. 1988. Groundwater quality; a data analysis protocol. *Journal of the Water Pollution Control Federation* 60:1938–1945.

9.4 MONITORING ACID LAKES IN QUÉBEC

(Prepared by M. Lachance, Univ. of Québec, J. Haemmerli, Environment Canada, and B. Bobée, Univ. of Quebéc, Ste-Foy, Québec, Canada)

The Québec area, located on the precambrian shield, has often been identified as one of the most sensitive regions to acidification (Altshuller and Mc Bean, 1979; 1980; Harvey et al. 1981; Shilts et al. 1981; Lachance et al. 1985). Lakes surveys done throughout Québec have shown some evidence of acidification in the aquatic ecosystems by acid precipitation (Lachance et al. 1985; Dupont and Grimard, 1986; Dupont, 1988). A close relationship has been observed between the sulfate concentration in the precipitation and lakewaters, which indicates how important atmospheric transport is on the chemical composition of the surface waters.

Because of the threat to aquatic ecosystems from this pollution problem and the fear of the long term consequences of an increase or a reduction in emissions of the pollutants that cause acidity, Environment Canada has established a monitoring network to detect trends in the acidification of surface waters in the province of Québec (Haemmerli et al. 1984).

The spatial aspects of this monitoring network were planned using the analysis of the data base from a survey conducted by Environment Canada during the winter of 1982 on 158 lakes (Bobée et al. 1983). The sampled region is located on the Canadian Shield, north of the Saint Lawrence River, between the Ottawa River and the Saguenay River. The planning of the temporal aspects of the monitoring was based on theoretical considerations and on a summary analysis of an available time series on a calibrated watershed located in the region under study (Bobée et al. 1983). The effects of atmospheric pollutants on aquatic ecosystems have been under study ever since 1981 at the Lake Laflamme watershed located in the Boreal Forest (Forêt Montmorency) 80 km north of Québec city (Papineau, 1983). Weekly samples have been taken in order to monitor the temporal variability of the quality of the lake water.

This case study reconstructs the steps used to create and implement a network to detect acidification trends. The actual steps in the process we set up were based on highly advanced statistical methods and on supplementary data not available at the time Environment Canada established its network. More specifically, this case study demonstrates how an adequate use of available information can lead to a rational and optimal design of a sampling plan whose purpose is the study of both the spatial and temporal aspects of an environmental problem.

9.4.1 General Concepts

The different steps used in the acquisition of water quality data follow the same objective process, regardless of the difficulty of the problem to be solved (Wilson, 1974; Sanders et al. 1987; Sherwani and Moreau, 1975; Bobée and Sasseville, 1978). Because of the complexity of natural aquatic systems and the difficulty in evaluating changes resulting from anthropogenic activities, which are sometimes very subtle, special attention must be paid to the planning of each step in the creation of a data acquisition network. The best procedure when gathering information is to have well-defined objectives. That is, they must be clear enough to allow for the conception of an adequate sampling plan and an optimal acquisition of information, subject of course, to available resources. The many steps in the planning of a water quality data acquisition network are illustrated on Figure 9.22.

As shown in Figure 9.22, the planning activities include several steps ranging from the definition of the objectives up to the use of the informa-

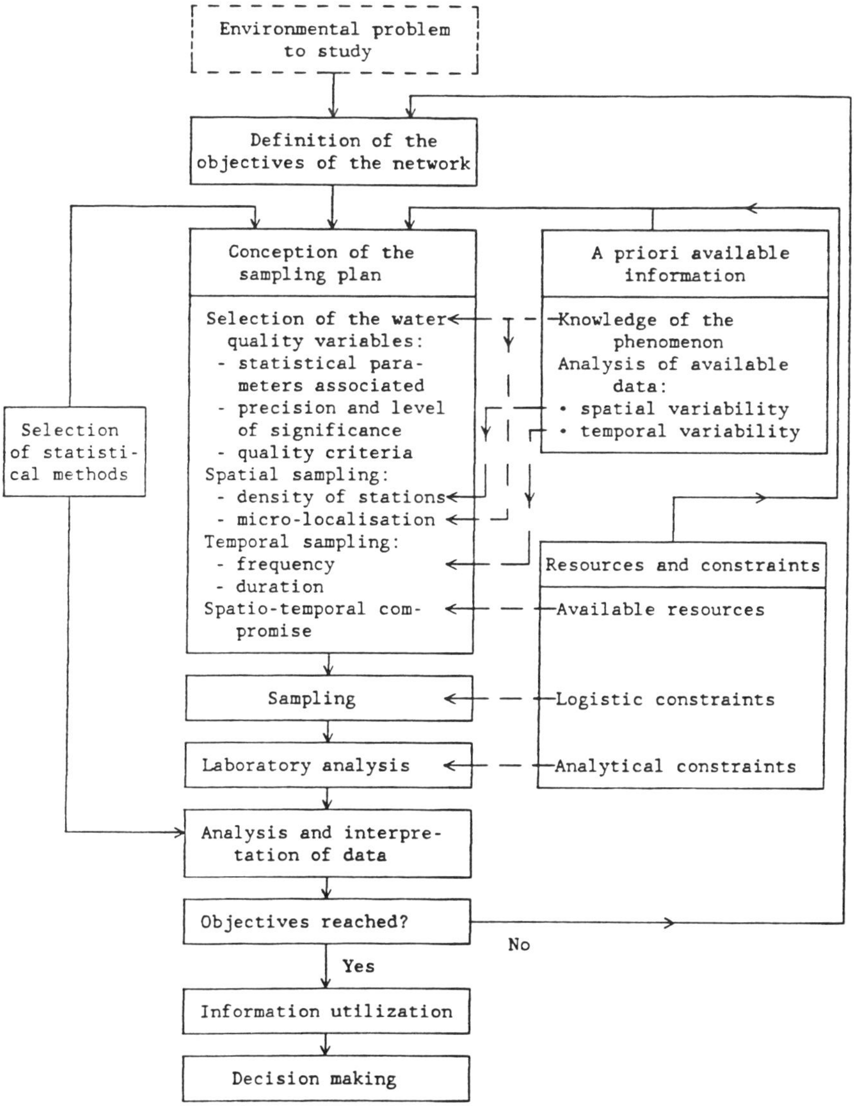

Figure 9.22 Diagram illustrating the different steps in the design of a data acquisition network (from Haemmerli et al. 1984).

tion. Our study deals only with the essential steps that include the use of available information and the consideration of available resources and existing constraints in the design of a sampling plan.

The detection of changes in the acidification of surface waters must be possible under a range of acid atmospheric inputs and take into account the acid buffering capacity of the watersheds. Consequently, the trend detection operation includes both a temporal and spatial analysis (improvement or degradation of water quality at a specific site and extension or recession of affected areas). The two components of this analysis can conflict when available resources are limited, as time trends detection implies a high frequency sampling at a limited number of stations. On the other hand, spatial trend detection requires a high density network to monitor the spatial variability and to control the regional representativity of temporal stations.

Since these two components are in conflict, two distinct sampling plans are needed: one for spatial aspects and the other for temporal aspects. The planning of these two phases is essential and cannot be realized without a knowledge of the spatio-temporal variability of the water quality parameters selected to meet the objectives of the network.

9.4.2 Spatial Variability Analysis

An analysis of the spatial variability is made using available data in order to subdivide the future network area into a number of homogeneous zones called strata. These strata are determined on the basis of their homogeneity in relation to a set of water quality variables selected to meet the network objectives (that means trend detection in acidification).

The water quality variables considered for this spatial analysis are pH (measure of water acidity), alkalinity (indicator of the buffering capacity of acid inputs), sulfate concentration (indicator of the aggression level by atmospheric sulfur inputs) and the sum of calcium and magnesium concentrations (characterizing the global mineralization). These four chemical variables were selected because they offer high-information content on the sensitivity or the acidity level of the lakes.

The multivariate methods used were correspondence analysis (CA) and ascending hierarchical classification (AHC) (Lebart et al. 1984; Jambu and Lebeaux, 1983). The combined use of these two statistical techniques has led to many applications in water quality data analysis (Lachance, 1988; Bobée and Lachance, 1984; Lachance and Bobée, 1982; Lachance et al. 1979).

Correspondence analysis was developed principally for the analysis of contingency tables analysis and cannot be directly applied to the original data. A binary coding of data is then used. Let X be a $(I \times Q)$ table with I observations (stations) as lines and Q variables as columns. On each variable q, we define J_q categories (variation intervals), each one to include a number of observations. Each observation X_{iq} is coded

1 or 0 depending on its presence in or absence from the category. The new coded table k now has I lines corresponding to I observations and J (ΣJ_q) columns corresponding to Q variables separated in J_q categories.

For the present application, we decided to define for each variable five categories with an approximately equal number of observations ($J_q = 5$; $q = 1, \ldots, 4$). Therefore the analysed table has 158 lines corresponding to the 158 surveyed lakes and 20 columns corresponding to four variables, each one separated into five categories.

With this application of CA, it is possible to explain 20.1 and 15.1% of the total variability on the first two principal axes. Inspection of the position of the "category points" on the plane formed by axes 1–2 (Figure 9.23*a*) shows that pH, alkalinity and Ca + Mg have a similar variation, whereas the sulfate variable behaves differently. Indeed, when polygonal lines are drawn to join the ordered items, we observe a similar course for these three variables. As for sulfate, the different course is caused mainly by the proximity of low sulfate values to the intermediate values of the three other variables.

Points representing the categories defined for alkalinity, pH and Ca + Mg are distributed as a parabola. It should be remembered that this type of course, called Guttman effect, is obtained when the phenomenon is mainly unidimensional (Benzécri and colleagues, 1976). This type of representation indicates that water mineralization, described chiefly by alkalinity, pH and Ca + Mg, is a space continuum phenomenon of the entire region studied.

By means of AHC, five classes of lakes were determined using the distances calculated between the coordinates of the points on the first eight factorial axes. Figure 9.23*b* shows the projection of the centres of gravity of the five classes on the same graph with the points representing the categories defined for the variables. In order to be able to picture the dispersion of the cloud corresponding to each of the five classes of lakes (see Figure 9.23*b*), the groups of points are shown using inertia ellipses, calculated according to Grelet and Lebeaux (1980). The size of these ellipses shows that the dispersion of each of these clouds is comparable. The absence of superposition confirmed that these five classes of lakes constitute quite distinct groups.

The results of the classification are arrived at by examining proximities between "category points" and the centers of gravity of the classes of lakes. According to the barycentric property of CA, the distance between a category point and a class of observations characterizes the degree of importance of this point in the formation of the class. Inspecting Figure 9.23*b*, one sees that class E is characterized by high sulfate values

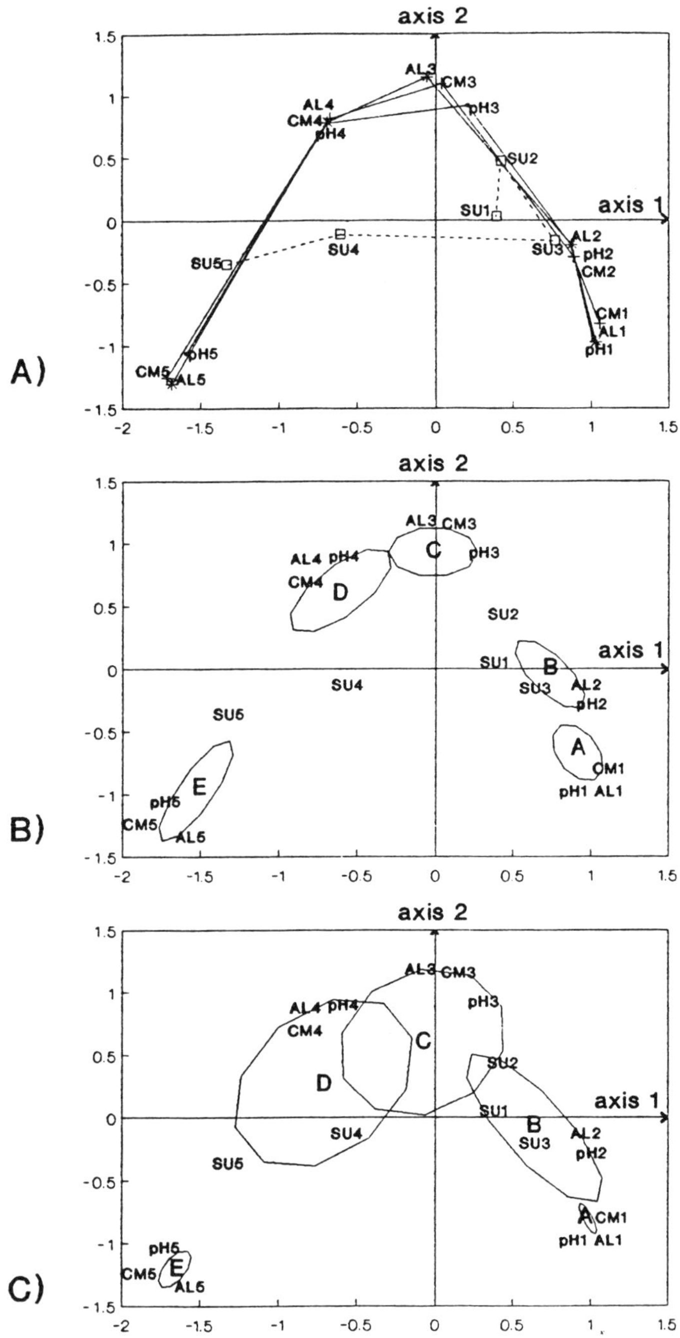

Figure 9.23 Representation in the plane formed by the two first principal axes of: a) the "category points"; b) the classification of "lake points" without constraint; and c) the classification of "lake points" with a contiguity constraint.

(SU5). It is also possible to characterize lake classes more objectively by calculating the contribution of each category point to the position of the center of gravity of the class in the factorial space. Details of this procedure are given by Lachance (1988).

By examining the information in Figure 9.23*b*, it is clear that the points characterizing each of the five classes are the categories defined for the alkalinity, Ca + Mg and pH. Except for point SU5 which characterizes class *E*, sulfate is a less determinant variable.

By reporting the class of each lake on the map of the region being studied, we can check whether or not these five classes delineate geographic zones. During the first step of this study, five geographic zones were identified using this method and the supplementary information on the physiographic characteristics of the territory (Bobée et al. 1983). This method is particularly precarious when we find neighboring lakes belonging to different classes in the same area.

In order to obtain more objective zoning, a contiguity factor was introduced as part of the classification method. With this technique (Lebart, 1978), stations are automatically partitioned into geographically neighboring classes; these groupings are still made on the basis of similar characteristics for the chemical variables considered. Therefore, to use this method, a contiguity matrix has to be prepared giving the most neighboring points for each lake. Thiessen polygons are then drawn for each lake (Lachance, 1988). Later, two contiguity notions were defined. A lake is identified as contiguous to another lake when the two have either a common polygon side or a common polygon vertex or side. The latter classification was retained here (Figure 9.23*c*).

By using a contiguity constraint, there is an important dispersion increase for classes *B*, *C*, and *D*, and a decrease for classes *A* and *E*. The larger intra-class dispersion for *B*, *C*, and *D* can be explained by the fact that this method sometimes requires that lakes with very different characteristics be grouped together because of their neighborhood. The five geographic zones obtained with this method are shown in Figure 9.24. The new zones correspond roughly to the zones delineated by Bobée et al. (1983). However, they are now obtained in a more objective fashion.

9.4.3 Spatial Optimization

The planning of the spatial sampling essentially consisted in determining a density of stations per stratum in the region. To determine this density,

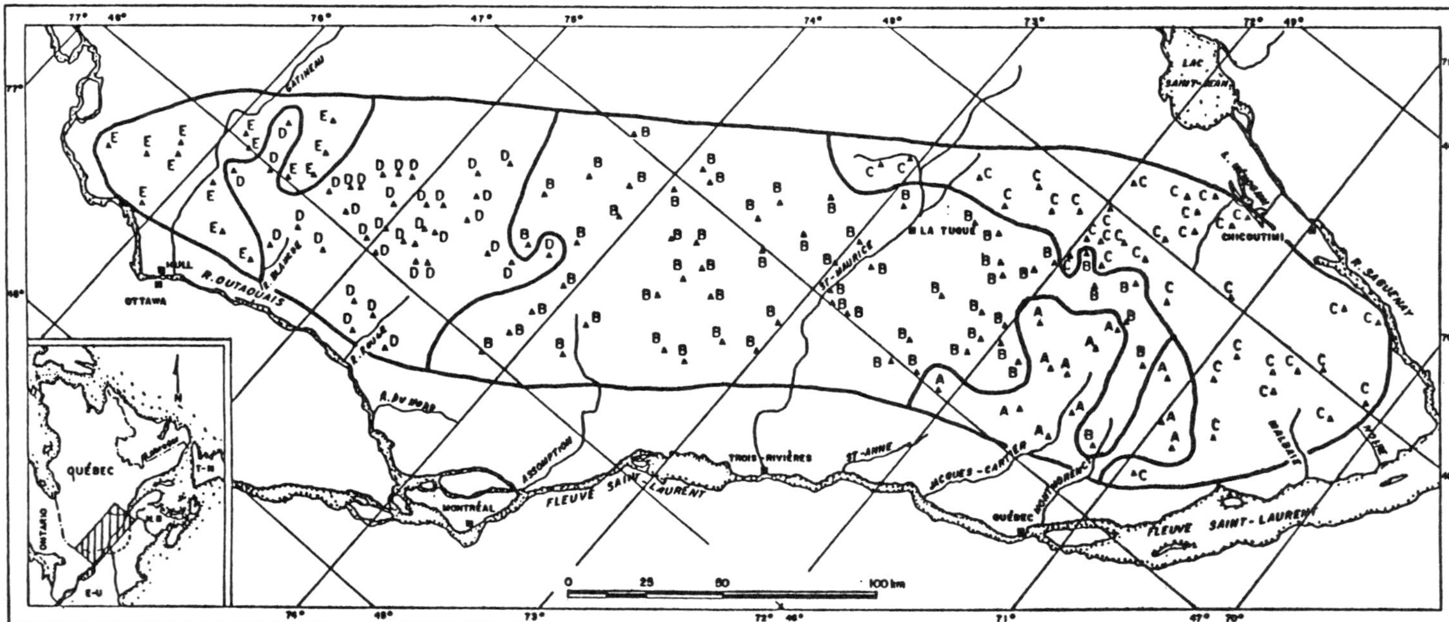

Figure 9.24 Studied region with the five geographical zones delineated by the application of hierarchical classifications with a contiguity constraint.

we used stratified random sampling with optimum allocation. This method has the advantage of taking into account, for each stratum, the variance of the variable considered, the number of lakes and the sampling cost. If the hypothesis is that the unit sampling price is the same for each stratum, the number n_i of stations per stratum i is given by the following equation (Cochran, 1977):

$$n_i = n \frac{N_i\, s_i}{\sum_{i=1}^{5} N_i\, s_i} \qquad (9\text{-}3)$$

where

n = total number of stations to be allocated among the different strata;
N_i = population of stratum i;
s_i = standard deviation of the variable for the stratum i.

Table 9.2 shows the results of the application of this equation for each of the four chemical variables: alkalinity, Ca + Mg, pH, and SO_4. We thus obtain four spatial sampling plans (one for each variable). A compromise may be obtained by calculating the arithmetic average of the number of stations for each stratum. This value is not optimal for each variable but is the best acceptable according to the objectives.

Table 9.2. Sampling of spatial variability: number of lakes, standard deviation, and number of lakes to sample.

PARAMETER		STRATUM 1	2	3	4	5
NUMBER OF LAKES PER STRATUM (N_i)		104	614	217	176	81
Standard deviation[1] (s_1)	alkalinity	9.0	41.5	95.6	166.3	572.9
	Ca + Mg	15.1	42.6	74.8	164.9	534.5
	pH	0.28	0.40	0.38	0.52	0.42
	sulfate	17.7	19.9	12.6	30.2	73.5
Number of lakes[2] to sample (n_i)	alkalinity	1	20	17	24	38
	Ca + Mg	2	22	14	25	37
	pH	6	51	17	19	7
	sulfate	7	43	10	19	21
	compromise	4	34	14	22	26

[1] Units are given in μeq /L except for pH which is given in units of pH.
[2] Total number of lakes to sample is fixed at 100.

9.4.4 Temporal Variability Analysis

To determine the sampling frequency and the duration of the operation of a network needs to have an "a priori" knowledge of temporal variability. Available data for temporal variability analysis came from Lake Laflamme, located inside the studied region (latitude and longitude of 47°19′ and 71°07′ respectively). Weekly measurements of physico-chemical variables were available for the lake for a period of eight years (1981–1988). This lake meets all the selection criteria used for the survey of the lakes in the region. Because of the physico-chemical characteristics retained, this lake may be seen as belonging to zone *A* (Figure 9.24).

For the analysis of the time aspects, we continued to consider the four following physico-chemical variables: alkalinity, sulfate, Ca + Mg and pH. The temporal evolution (for the period 1982–1987) of these four variables is illustrated in Figure 9.25. The alkalinity and the sum of Ca + Mg concentrations show very pronounced cyclic variations related to the annual hydrological cycle. There is a sudden decrease in the spring and a gradual increase during the summer and autumn. Sulfate and pH follow an increasing trend, whereas alkalinity, in spite of an important seasonal variation, exhibits a decreasing trend.

9.4.5 Temporal Optimization

Since the main objective for this network is to determine a temporal trend, temporal optimization may be seen as determining in advance the sample size needed to apply statistical tests in the future. The hypotheses to be verified are the following:

H_0: there is no trend
H_A: there is a trend

To verify these hypotheses, the authors use the relation between the power of a Student test, the detected trend level and the desired level of significance (Lettenmaier, 1976):

$$1 - \beta = F_g\,(N_T - Z_{1-\alpha/2}) \tag{9-4}$$

where

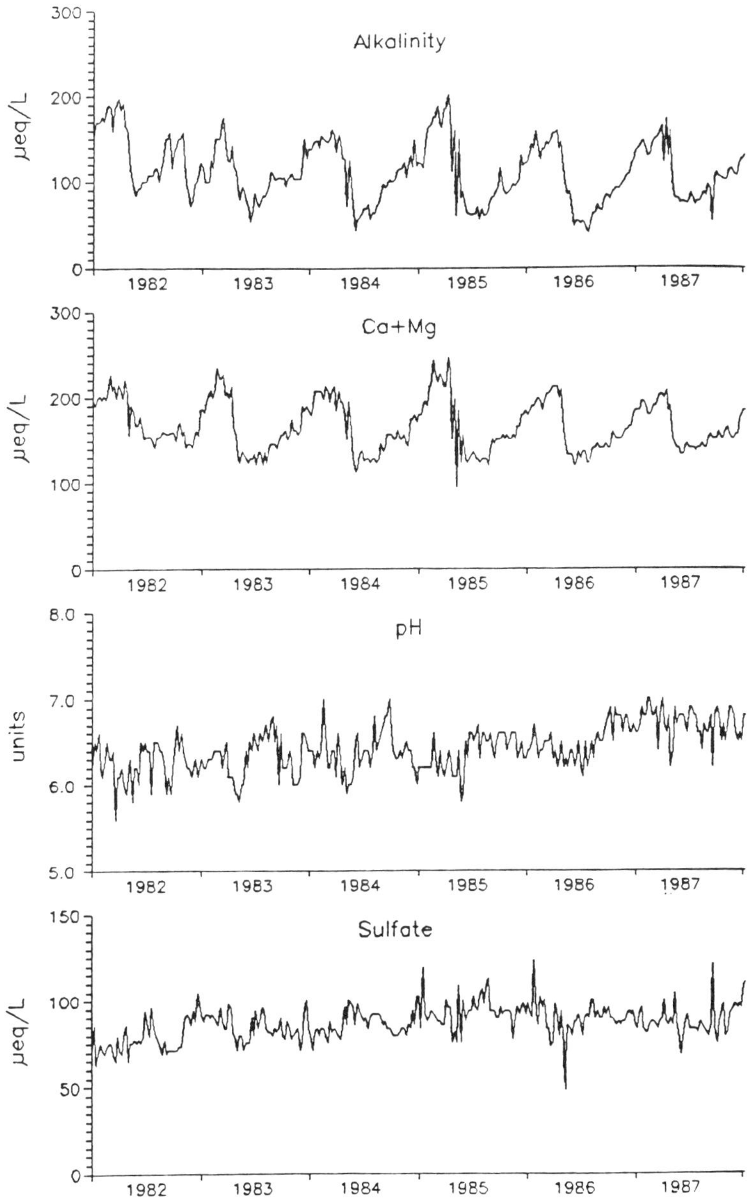

Figure 9.25 Temporal evolution of alkalinity, Ca + Mg, pH, and sulfate on Lake Leflamme during 1982–1987 period.

F_g is the cumulative function of the standardized normal variate;
$Z_{1-\alpha/2}$ is the quantile of the normal distribution for a probability of non exceedence $1 - \alpha/2$;
N_T is a number (dimensionless) which takes into account the trend level, the length of the series and is defined according to the kind of the trend considered.

Supposing that the trend, if it happens, will be linear instead of in steps, the expression of N_T for a linear trend is the following (Lettenmaier, 1976):

$$N_T = \frac{Tr}{\sigma_\epsilon} \frac{\sqrt{n^*}}{\sqrt{12}} \qquad (9\text{-}5)$$

where

Tr is the mean level increase during a period of time;
σ_ϵ is the standard deviation of the random component of the series; n^* is the number of independant samples.

If the power of the test is fixed $(1 - \beta)$ at 0.9, the relation 9-4 is simplified to:

$$N_T - Z_{1-\alpha/2} = 1.282$$

If N_T is replaced by the expression given by equation 9-5, it becomes an equation relating the sample size n^* to a trend level Tr/σ_ϵ and to a significance level α when the power of the test is fixed:

$$n^* = \frac{12\,(1.282 + Z_{1-\alpha/2})^2}{(Tr/\sigma_\epsilon)^2} \qquad (9\text{-}6)$$

When the sample size is lower than 30, it is better to replace the quantile $Z_{1-\alpha/2}$ of the normal distribution by the quantile $t_{1-\alpha/2;n}$ of the Student distribution.

The preceding equations are valid only for independent samples. In practice, one is often faced with dependent samples in a series; this dependency reduces the information content of the series. The information content of a stationary series of n dependent observations can be calculated by the following relation (Lettenmaier, 1976; Sanders et al. 1987):

$$\frac{1}{n^*} = \frac{1}{n} + \frac{2}{n^2}\left[\sum_{p=1}^{n-1} (n-p)r_p\right] \tag{9-7}$$

where

p = interval of time between successive observations;
n = total number of observations in the series;
n^* = effective number of observations;
r_p = autocorrelation coefficient for lag p.

The following procedure was then adopted: for each physico-chemical variable, the autocorrelation function was calculated on series made stationary and then n^* was calculated for different values of n and r_p. To do this, the normality had to be verified for the distributions. For each of the variables, the normality of the distributions was accepted despite some deviations in the normality for alkalinity and Ca + Mg (Lachance, 1988).

To obtain stationary series, a seasonal component and a linear trend component have to be subtracted. For the extraction of a seasonal component on each data series (between January 1, 1982 and December 31, 1987), the moving average of the 26 preceding weeks and of the 26 following weeks was calculated. This computation was made possible because data were available for the six last months of 1981 and the six first months of 1988. We then obtained the seasonal component by calculating the inter-annual average of the deviations between observed value and moving average for each of the 52 weeks of the year. A deseasonalized series is obtained by subtracting the seasonal component from the initial series.

For each of the deseasonalized series, the linear trend was calculated by the least squares method. From the random component of the linear regression equation (Lachance, 1988), the authors have the stationary series on which the autocorrelation function is calculated for the lags 1 to 30 (Figure 9.26). The standard deviation of the autocorrelation coeffi-

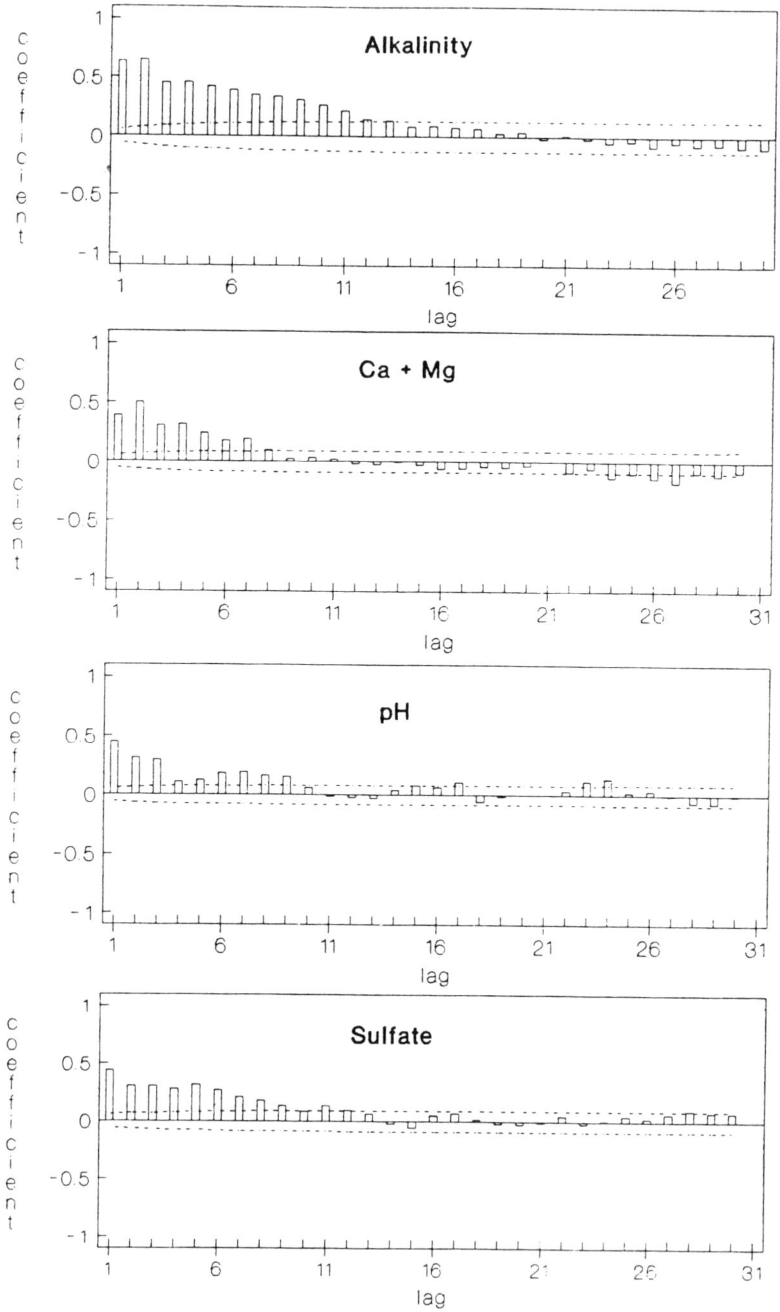

Figure 9.26 Autocorrelation function on stationary series of alkalinity, Ca + Mg, pH, and sulfate.

cients gives an estimate of the significance level of the latter. One can see that the persistence is significant over a relatively long period of time, notably for alkalinity (13 weeks), sulfate (12 weeks), pH (9 weeks) and Ca + Mg (7 weeks).

For a given period of time, the effective number n^* of observations was calculated according to different numbers n of samples from Eq. 9–7. The autocorrelation coefficients r_p used in this equation are the p positive non-zero coefficients calculated from the stationary series. A smoothing of the function (moving average of 5 terms) was performed in order to correct the coefficients for random fluctuations.

For a weekly sampling frequency ($n = 52$), the effective number n^* of independent observations per year for each of the physico-chemical variables is as follows:

Alkalinity	Ca + Mg	pH	Sulfate
5.1	10.0	10.6	8.4

This effective number (n^*) of observations may be considered as a good approximation of the maximum number (n^*_{max}) of independent observations per year for each of the variables. Indeed it was shown (Lachance, 1988) that the increase in the effective independent samples flattens when the number of samples is superior to 40 per year.

The quantity n^*/n^*_{max} may be considered as a measure of the saturation level of information. Therefore, for each variable considered, the annual number of samples needed for different information saturation levels is calculated (Table 9.3). A compromise may be obtained by calculating

Table 9.3. Number of samples per year according to the sampling saturation level.

SATURATION LEVEL n^*/n^*_{max} (%)	ALKALINITY	Ca + Mg	pH	SULFATE	COMPROMISE
50	3	5	7	5	5
60	4	8	8	7	7
70	5	11	11	10	9
80	7	17	15	14	13
90	13	28	25	23	23

the arithmetic average (or weighted average depending on the variables) of annual sampling frequencies of each variable. For a given information saturation level fixed at 80% (reasonable saturation level), an annual sampling frequency of 13 would be needed.

Knowing n^* for different sampling frequencies, we can deduce the total sampling period of time to detect trend to standard deviation ratios Tr/σ_ϵ equal to values such as 2.0, 1.5, 1.0, and 0.7 (Table 9.4). These results show that to detect a trend level Tr/σ_ϵ lower than 1.5 before a reasonable period of approximately 10 years may not be possible. These results also show that it is not desirable to increase the annual sampling frequency from 13 to 26, as it will not sufficiently reduce the sampling period.

From all the results obtained by the analysis of temporal variability, we can conclude that a monthly sampling frequency is optimal when the high level of persistence in data and the acceptable trend detection level for a 10-year sampling period are taken into account.

Table 9.4. Total sampling period (in years) needed to detect some trend levels according to different sampling frequencies for a power of the test fixed at 90% and a significance level fixed at 5%.

Tr/σ_ϵ	ANNUAL SAMPLES	ALKALINITY	Ca + Mg	pH	SULFATE
2.0	6	8.3	6.0	6.3	6.7
	8	7.4	5.2	5.0	5.8
	13	6.8	4.3	4.0	4.7
	26	6.4	3.6	3.3	4.0
1.5	6	14.8	10.7	11.3	12.0
	8	13.2	9.2	9.0	10.4
	13	12.1	7.6	7.1	8.5
	26	11.3	6.3	5.8	7.2
1.0	6	33.4	24.0	25.4	26.9
	8	29.7	20.6	20.2	23.3
	13	27.3	17.0	15.9	19.0
	26	25.4	14.3	13.1	16.2
0.7	6	68.1	49.0	51.8	54.9
	8	60.7	42.1	41.2	47.6
	13	55.7	34.7	32.4	38.8
	26	51.9	29.1	26.7	33.0

9.4.6 Spatio-Temporal Compromise

To reach the objective of knowledge of spatial variability, a great number of spatial stations, sampled at a low frequency, have to be allocated. Also, to meet the other objective, which is the knowledge of temporal variability, a reduced number of temporal stations sampled at a high frequency had to be provided. Limited resources generally limit the total number of samples per year to be analyzed. To meet both objectives when resources are limited, a sampling plan was drawn up as a result of a spatio-temporal compromise.

The total number of samples N_0 per year fixed by available resources may be expressed as follows:

$$N_0 = K\,t + L\,s \tag{9-8}$$

where

K = number of temporal stations;
L = number of spatial stations;
t = temporal stations sampling frequency;
s = spatial stations sampling frequency;

If spatial tours are synchronized with visits of temporal stations, the latter can also contribute in providing spatial information. The effective number of spatial stations thus becomes equal to the total number of stations $M = K + L$.

The total number of samples N_0 per year may be expressed as a function of K and M:

$$N_0 = K\,(t - s) + M\,s \tag{9-9}$$

If N_0 is fixed (say 300), it is possible to visualize graphically the relationship between K and M for different values of t (6, 8, 13, or 26) and s (1 or 2) (see Figure 9.27). The graph can be used as a nomogram that relates the allocation of temporal stations K with the total number of stations M leaving both temporal and spatial sampling frequency fixed.

A quantity π may be defined to express the relative weight of temporal aspects to spatial aspects:

$$\pi = \left(\frac{K\,t}{M\,s}\right) \tag{9-10}$$

Points corresponding to π = 1, 2, and 4 have been reported in Figure 9.27. One can see that this quantity π can be used as a guide when selecting temporal and spatial sampling plan scenarios.

If a reasonable limit for π is fixed, say between 1 and 5, and if a frequency of 13 samples per year is chosen, 14 spatio-temporal sampling plan scenarios can be built, from a total number of samples N_0 = 300, including 9 scenarios with one spatial tour per year (s = 1) and 5 scenarios with s = 2 (Table 9.5).

To allocate the M spatial stations to the five homogeneous zones, the authors used the random stratified sampling with optimum allocation

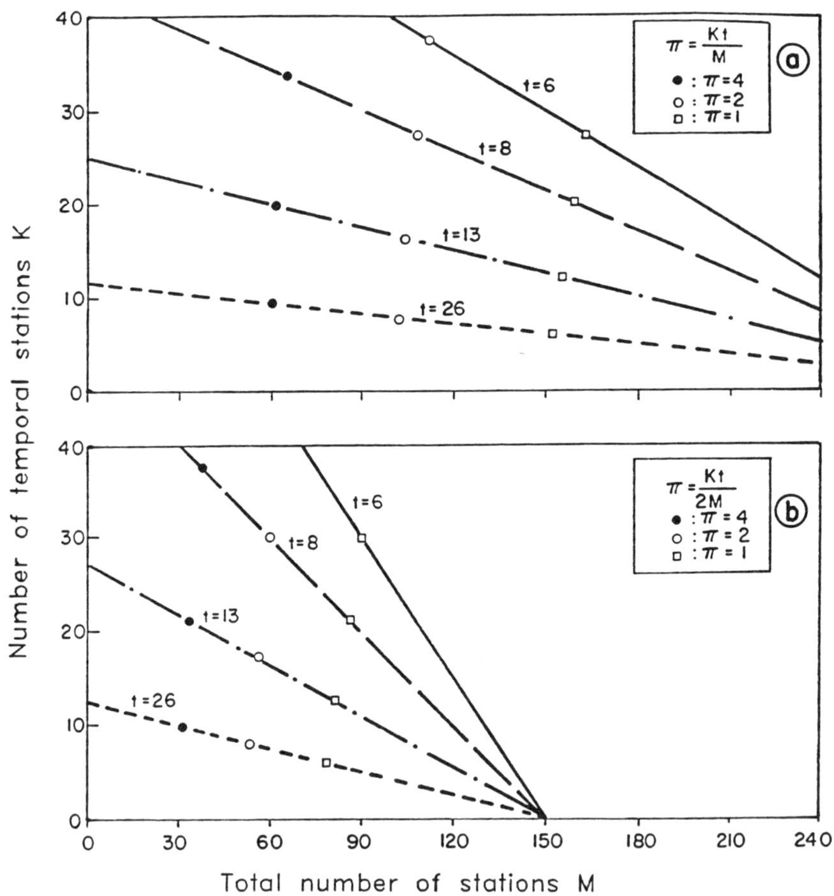

Figure 9.27 Nomogram usable to get the number of temporal stations in relation to the total number of stations when the sampling frequency is fixed.

Table 9.5. Sampling plan scenarios giving the allocation, per stratum, of the number of spatial and temporal stations on the basis of 300 samples per year and a frequency of 13 samples per year.

SCENARIO	s	K	M	π	K_1	K_2	K_3	K_4	K_5	M_1	M_2	M_3	M_4	M_5
1	1	20	60	4.33	2	10	3	3	2	4	20	8	13	15
2	1	19	72	3.43	2	9	3	3	2	4	24	10	16	18
3	1	18	84	2.79	2	9	3	2	2	4	28	12	18	22
4	1	17	96	2.30	2	8	3	2	2	4	33	13	21	25
5	1	16	108	1.93	2	8	2	2	2	4	37	15	24	28
6	1	15	120	1.62	2	7	2	2	2	5	41	17	26	31
7	1	14	132	1.38	2	6	2	2	2	5	45	19	29	34
8	1	13	144	1.17	2	5	2	2	2	6	49	20	32	37
9	1	12	156	1.00	2	4	2	2	2	6	53	22	34	41
10	2	22	29	4.93	2	11	4	3	2	4	9	4	5	7
11	2	20	40	3.25	2	10	3	3	2	4	13	5	8	10
12	2	18	51	2.29	2	9	3	2	2	4	17	6	11	13
13	2	16	62	1.68	2	8	2	2	2	4	21	8	13	16
14	2	14	73	1.25	2	6	2	2	2	4	24	10	16	19

Constraints for $M_i \geq 4$ and $K_i \geq 2$ for i = 1 to 5;
s: number of spatial tours per year;
K: number of temporal stations; K_i (i = 1 to 5) = number per stratum;
M: total number of stations (spatial and temporal); M_i (i = 1 to 5) = number per stratum;
π: relative weight of temporal aspects on spatial aspects.

method. However, to allocate the K temporal stations, they preferred to use the proportional random allocation method because of the lack of temporal information on the entire region being studied. Because of limited resources, limits had to be placed on the minimum number of stations per stratum. For the planned network, the authors provided a minimum number of temporal stations per stratum: $K_i \geq 2$. In fact, just one station per zone would be insufficient to adequately control the representativity of stations. Also provided was a minimum number of stations ($M_i \geq 4$) for the spatial sampling plan.

The selection of two spatial sampling tours considerably reduces the number of spatial stations for a given weight π. If only one spatial tour is carried out, it should then be done during a season with stable hydrological conditions, like winter. A second tour during another season allows us to ascertain the representativity of the sampling season and to make sure that spatial variability does not result from a seasonal effect. However, in the case where temporal stations are available and allow for this hypothesis to be verified, scenarios with two spatial tours should be eliminated when choosing a spatio-temporal compromise. A choice between different scenarios finally depends on the importance of the temporal or spatial aspects to be favored. Nevertheless, the final choice depends on the decision of the government agency responsible for establishing the network.

9.4.7 Conclusion

The authors have drawn up a methodology to develop a sampling plan to study an environmental problem with temporal and spatial aspects. This methodology was applied here to the design of a trend detection network in the acidification of lakes. A multivariate analysis of spatial information has led to the determination of homogeneous zones that allowed for the application of stratified sampling for spatial aspects. The analysis of the temporal information content of a long series of data, available on a lake of the region, demonstrated that a monthly sampling frequency was essential if a trend is to be detected in a reasonable period (say 10 years). Sampling plan scenarios, resulting from a spatio-temporal compromise, were elaborated in order to maximize the acquisition of information, while constraints imposed by available resources were taken into account.

The methodology presented here was developed using two steps. The first step, described in detail by Bobée et al. (1983), has already led to the design and establishment of a sampling network operated by Environment Canada (Haemmerli et al. 1984) to detect trends in the acidification of Québec surface waters. This network, which began operation in 1983, was initially composed of 35 stations used to study spatial aspects, 10 of which were sampled every two months for the study of temporal aspects. As one can see, this sampling plan is somewhat different from the sampling scenarios proposed in this study, but was developed with a similar procedure using data available at that time.

The methodology described here for the design of such a network can be used for wider regions such as Eastern Canada, and can also be applied to the study of other environmental problems.

REFERENCES

Altshuller, A.P. and G.A. McBean (1979). "The LRTAP problem in North America: a preliminary overview." Report of the United States-Canada Research Consultation Group on the Long-Range Transport of Air Pollutants. U.S.A. State Department and Canada Department of External Affairs, 48 p.

Altshuller, A.P. and G.A. McBean (1980). Second Report of the United States-Canada

Research Consultation Group on the Long-Range Transport of Air Pollutants. U.S.A. State Department and Canada Department of External Affairs, 39 pp.

Benzécri, J.P. and Coll. (1976). *L'analyse des données. Tome II: l'analyse des correspondances.* Dunod, Paris, 2nd edition, 619 pp.

Bobée, B. and M. Lachance (1984). "Multivariate Analysis of Parameters Related to Lake Acidification in Quebec." Water Resour. Bull., 20:545–556.

Bobée, B., M. Lachance, J. Haemmerli, A. Tessier, J.Y. Charette and J. Kramer (1983). Évaluation de la sensibilité à l'acidification des lacs du sud du Québec et incidences sur le réseau d'acquisition de données. INRS-Eau, rapport scientifique no. 157, 198 pp, 4 annexes, 1 appendice.

Bobée, B. and J.L. Sasseville (1978). Approche rationnelle aux problémes environnementaux. Eau du Québec, 11:189–197.

Cochran, W.G. (1977). *Sampling Techniques,* 3d edition, John Wiley & Sons, 428 pp.

Dupont, J. (1988). État de l'acidité des lacs de la région hydrographique de l'Outaouais. Environnement Québec, direction de la qualité du milieu aquatique, publication PA-29, 99 pp.

Dupont, J. and Y. Grimard (1986). Systematic Study of Lake Water Acidity in Québec. Water Air Soil Pollut., 31:223–230.

Grelet, Y. and M.O. Lebeaux (1980). Lecture commentée de sorties de programmes en analyse de données. Bulletin no. 11 de l'A.D.D.A.D. (Tour 45–55-E2, 4, Place Jussieu, 75230 Paris cedex 05).

Harvey, H.H., P.T. Dillon, J.R. Kramer, R.C. Pierce and D.M. Whelpdale (1981). "Acidification in the Canadian Environment: Scientific Criteria for Assessing the Effects of Acidic Deposition on Aquatic Ecosystems." Associate Committee on Scientific Criteria for Environmental Quality, CNRC no. 18475, Ottawa, 369 pp.

Haemmerli, J., B. Bobée and M. Lachance (1984). Élaboration d'un réseau de détection de tendances dans l'acidification des eaux de surface du Québec. Symposium canadien d'Hydrologie (Laval University, Québec, June 10–12).

Jambu, M. and M.O. Lebeaux (1983). Cluster Analysis and Data Analysis. Elsevier—North Holland, Amsterdam, 898 pp.

Lachance, M. (1988). Élaboration d'une méthodologie d'évaluation de la qualité de l'eau: application au problème de l'acidification des eaux lacustres du Québec. Ecole nationale Supérieure desMines de Paris, doctorate thesis.

Lachance, M., B. Bobée and D. Gouin (1979). Characterisation of the Water Quality in the Saint Lawrence River: Determination of Homogeneous Zones by Correspondence Analysis. Water Resour. Res., 15:1451–1462.

Lachance, M. and B. Bobée (1982). Methodology for the Characterization of Water Quality: Analysis of Time-Dependent Variability. J. Hydrol., 59:275–291.

Lachance, M., B. Bobée and Y. Grimard (1985). Sensitivity of Southern Quebec Lakes to Acidic Precipitation. Water Air Soil Pollut., 25:115–132.

Lebart, L. (1978). Programme d'aggrégation avec contraintes. Cahiers de l'Analyse des Données, 3:275–287.

Lebart, L.A. Morineau and K.M. Warwick (1984). Multivariate Descriptive Statistical Analysis. John Wiley & Sons, New York, 231 p.

Lettenmaier, D.P. (1976). Detection of Trends in Water Quality Data from Records with Dependent Observations. Water Resour. Res., 12:1037–1046.

Papineau, M. (1983). Le bassin du lac Laflamme: caractérisation physico-chimique du lac et établissement du bilan ionique en relation avec le transport à grande distance des polluants atmosphériques. Environnement Canada, direction des eaux intérieures, région du Québec, 95 p.

Sanders, T.G., R.C. Ward, J.C. Loftis, T.D. Steele, D.D. Adrian and V. Yevjevich (1987). Design of Networks for Monitoring Water Quality. Water Resources Publications, Littleton, Colorado, 2nd ed., 328 p.

Sherwani, J.K. and D.M. Moreau (1975). Strategies for Water Quality Monitoring. Water Resour. Res. Inst., Univ. of North Carolina, Report no. 107.

Shilts, W.W., K.D. Card, W.H. Poole and B.V. Sandford (1981). Sensitivity of Bedrock to Acid Precipitation: Modification by Glacial Processes, Paper 81–14, Geological Survey of Canada, 7 p., 3 maps.

Wilson, A.L. (1974). The Design of Long-Term Programs for the Measurement of River Water Quality. Design of Environmental Information Systems, R.A. Deininger, Ann Arbor Science Pub., pp 199–226.

10
SUMMARY

An information systems approach to the design of water quality monitoring programs has been described. The approach emphasizes the need to carefully define the information expectations placed on the monitoring system and to use those expectations to guide specification of each component of the information system.

To organize the design approach, we first defined the term "water quality monitoring system." By following the flow of information through the various activities associated with water quality monitoring, it was possible to define the monitoring system as having six major components: (1) sampling; (2) laboratory analysis; (3) data handling; (4) data analysis; (5) reporting; and (6) information utilization. The design process specifies how each of these activities should be carried out if the information goals for the monitoring effort are to be achieved.

Before discussing the design process, we investigated the role of statistics in water quality monitoring. Each water quality sample is only a very small part of the total population, and data from each sample are combined to make inferences about the total water quality population. The use of statistics permits us to quantify the uncertainty in these inferences as we develop knowledge about water quality behavior in the environment.

With an appreciation for the role of statistics, an approach to design was recommended, involving the following five steps:

Step 1. Define information needs of management
Step 2. Define information that can be produced by monitoring
Step 3. Design monitoring network
Step 4. Document data collection procedures
Step 5. Document information generation procedures

In performing these steps there is a need to develop a clear picture of the information expectations, hopefully based upon a consensus among information users. Developing information expectations involves carefully defining exactly how the data will be analyzed. From information expectations, it is possible to establish the network design—where samples will be taken, when they will be taken and what will be measured. Details of system operation will then be based upon the network design and information expectations. The details will include sample collection and analysis procedures, data management and analysis, and report generation. A final design report should document these details and trace the flow of information all the way to its ultimate use in decision making.

The system design check list presented in the following gives a quick overview of the entire process.

I. Definition of Information Expectations (Goals)

A. Has the water quality information needed to assist decision making within the total management program been defined and documented?

B. Have the limitations on the ability of water quality monitoring to produce information been identified?

C. Has an "information expectations" report been prepared and a consensus relative to expectations been developed between information users and monitoring system personnel?

- Has the term "water quality" been defined in such a way that the variables to be measured can be defined?
- Have sampling locations and sampling frequencies been related to the information goals?
- Have the graphical/statistical means to convert the data into information (ie a data analysis protocol) been developed and documented?
- Are there well defined reporting and information utilization procedures?

II. Definition and Documentation of Monitoring System Operations (Is there a design report that addresses each of the following operations?)

A. Sampling

- Sampling routes, equipment maintenance, employee training
- Field sampling and analysis procedures
- Sample processing/preservation and transportation
- Quality assurance

B. Laboratory Analysis

- Scheduling and operational procedures
- Laboratory analysis methods
- Quality control in the laboratory
- Data recording/logging

C. Data Handling
- Screening and verification of data
- Computer hardware
- Data base management software
- Data storage and retrieval procedures

D. Data Analysis
- Graphical/statistical procedures
- Data analysis protocol
- Data analysis software

E. Reporting
- Format and contents
- Frequency of publishing
- Distribution
- Audit of reporting system's effectiveness

F. Information Utilization
- Program managers/technical staff
- Program administrators
- Policy makers/board of directors
- Public
- Audit of utilization effectiveness

The most appropriate approach to design of any particular water quality monitoring system will vary depending upon a number of factors including the background and experience of the designer, time and resources available for the design, and management setting within which the monitoring will take place. This book has attempted to provide a general guide to assist designers as they develop their own approaches. The case studies presented in Chapter 9 illustrate the individuality of specific monitoring programs. Yet each is based on a systematic and scientific design.

The rather detailed approach to water quality monitoring and information system design described in this book requires much more investment in time and money than has been common in the past. No excuses are required for this large investment, and none are given here. As the old adage goes, "You get what you pay for!." If the time and effort are not invested up front, what you get is described by Ward et al. (1986) as the "data rich but information poor" syndrome. The authors trust that this book has provided a means of avoiding this syndrome in water quality monitoring.

REFERENCES

American Public Health Association—American Water Works Association—Water Pollution Control Federation. 1985. *Standard Methods for the Examination of Water and Wastewater,* 16th ed., American Public Health Association, Washington, D.C.

Alexander, R.B., T.L. Schertz, and D.J. Ohe. 1989. A computerized system for the detection and analysis of long-term trends in water-quality data. U.S. Geological Survey Open-File Report 89–409, Reston, Virginia.

Association of State and Interstate Water Pollution Control Administrators. 1983. Instructions and definitions for "STEP" report on progress in the clean water program. ASIWPCA, Washington, D.C.

Association of State and Interstate Water Pollution Control Administrators. 1984. America's Clean Water: The State's Evaluation of Progress 1972–1982. ASIWPCA, Washington, D.C.

Ball, R.O., and R.L. Church. 1980. Water quality indexing and scoring. J. Envr. Engr. Div., Am. Soc. Civil Engrs., Vol. 106(EE4):757-771, August.

Bayley, G.V. and J.M. Hammersley. 1946. The effective number of independent observations in an auto-correlated time series. *Journal Royal Statistical Society,* 8(1-B):184–197.

Beckers, C.V., S.G. Chamberlain and G.P. Grimsrud. 1972. Quantitative methods for preliminary design of water quality surveillance systems. Socioeconomic Environmental Studies Series Report No. EPA-R5–72–001, U.S. Environmental Protection Agency, Washington, D.C., November.

Beckers, C.V., and S.G. Chamberlain. 1974. Design of cost-effective water quality surveillance systems. Report EPA-600/5–74–004, U.S. Environmental Protection Agency, Washington, D.C., January.

Bell, H.F. 1989. Water quality information systems: past, present and future—an industrial perspective. *Proceedings of the International Symposium on the Design of Water Quality Information Systems,* Information Series No. 61, Colorado Water Resources Research Institute, Colorado State University, Fort Collins, Colorado, pages 22–28.

Beyer, W.H. 1968. *Handbook of Tables for Probability and Statistics,* 2d. ed. Cleveland, Ohio: The Chemical Rubber Co.

Bloem, D.M. and J.L. Glicker. 1989. The evolution of a water utility's water quality information system. *Proceedings of the International Symposium on the Design of Water Quality Information Systems,* Information Series No. 61, Colorado Water Resources Research Institute, Colorado State University, Fort Collins, Colorado, pages 29–36.

Bowker, A.H., and G.J. Lieberman. 1972. *Engineering Statistics,* 2nd ed. Englewood Cliffs, New Jersey: Prentice-Hall.

Box, G.E.P., and G.M. Jenkins. 1976. *Time Series Analysis: Forecasting and Control,* 2nd ed., San Francisco: Holden-Day.

Bradley, J.V. 1968. *Distribution-Free Statistical Tests.* Englewood Cliffs, New Jersey: Prentice-Hall.

Brooksbank, P. 1989. NAQUADAT, A Redesigned Data Base. *Proceedings of the International Symposium on the Design of Water Quality Information Systems,* Information Series No. 61, Colorado Water Resources Research Institute, Colorado State University, Fort Collins, Colorado, pages 251–262.

Bros, W.E., and B.C. Cowell. 1987. A technique for optimising sample size (replication). *Journal of Experimental Marine Ecology* 114:63–71.

Brown, R.M. and N.I. McClelland. 1970. A water quality index—do we dare? *Proceedings of the National Symposium on Data and Instrumentation for Water Quality Management,* Water Resources Center, University of Wisconsin, Madison, Wisconsin, pages 364–383.

Cairns, J., Jr. 1989. Biological monitoring of water quality. *Proceedings of the International Symposium on the Design of Water Quality Information Systems, Information Series No. 61,* Colorado Water Resources Research Institute, Colorado State University, Fort Collins, Colorado, pages 338–346.

Caulcutt, R., and R. Boddy. 1983. *Statistics for analytical chemists.* London: Chapman and Hall.

Chatterjee, S., and B. Price. 1973. *Regression Analysis by Example.* New York: Wiley.

Clark, R.N. 1950. Discussion of paper by Velz, 1958. *Sewage and Industrial Wastes.* 22(5):683–684.

Cochran, W.G., and G.M. Cox. 1957. *Experimental Designs,* 2nd ed. New York: Wiley.

Cohen, J. 1988. *Statistical Power Analysis for the Behavioral Sciences,* 2nd ed. Hillsdale, New Jersey: Lawrence Erlbaum Associates.

Cohen, P., W.M. Alley, and W.G. Wilber. 1988. National water-quality assessment: future directions of the U.S. Geological Survey. *Water Resources Bulletin* 24(6):1147–1151.

Committee on Science and Technology. 1983. National Environmental Monitoring. Hearings of the Natural Resources, Agriculture Research and Environment Subcommittee of the Committee on Science and Technology, U.S. House of Representatives, Ninety-eighth Congress, 1st session, May and June, U.S. Government Printing Office, Washington, D.C.

Conover, W.J. 1980. *Practical Nonparametric Statistics,* 2nd ed. New York: Wiley.

Conover, W.J. and R.L. Iman. 1981. Rank transformations as a bridge between parametric and nonparametric statistics. *The American Statistician* 35(3):124–133.

Council on Environmental Quality. 1980. *Final report of the interagency task force on environmental data and monitoring.* Available from NTIS, U.S. Department of Commerce, Springfield, Virginia.

D'Agostino, R.B., and M.A. Stephens. 1986. *Goodness of Fit Techniques.* New York: Marcel-Dekker.

Dunlop, W.J., J.F. McNabb, M.R. Scalf, and R.L. Cosby, 1977. Sampling for Organic Chemicals and Microorganisms in the Subsurface. Environmental Protection Technology Series, Report No. EPA-600/2–77–176, Robert S. Kerr Environmental Research Laboratory, Ada, Oklahoma, August.

Dunnette, D.A. 1979a. A geographically variable water quality index used in Oregon, *Journal Water Pollution Control Federation,* 51(1):53–61.

Dunnette, D.A. 1979b. Stream sampling optimization using a water quality index. Draft report to the Water Quality Division, Department of Environmental Quality, State of Oregon, P.O. Box 1760-, Portland, Oregon 97207, March.

Edgington, E.S. 1987. *Randomization Tests.* 2nd edition, New York: Marcel-Dekker.

Elliott, J.M. 1977. *Some Methods for the Statistical Analysis of Samples of Benthic Invertebrates.* Scientific Publication 25. Cumbria, England: Freshwater Biological Association.

Ellis, J.C. and R.F. Lacey. 1980. Sampling: defining the task and planning the scheme. *Water Pollution Control* (U.K.) 79:452–467.

Environment Canada. 1983. Sampling for Water Quality. Water Quality Branch, Inland Waters Directorate, Ottawa, Ontario, Canada.

Everett, L.G. and K.D. Schmidt (Eds.) 1978. Establishment of water quality monitoring programs. Proceedings of a Symposium held in San Francisco, California, June 12–14, 1978, published by the American Water Resources Association, Minneapolis, Minnesota.

Ferris, C.D., F.E. Grubbs, and C.L. Weaver. 1946. Operating characteristics for the common statistical tests of significance. *Annals of Mathematical Statistics* 17:178–197.

Ficken, J.H. 1989. Advances in instrumentation for automatically monitoring water quality. U.S. Geological Survey Open-File Report 89–409, Reston, Virginia.

Fisher, R.A. 1954. *Statistical Methods for Research Workers.* 12th ed. New York: Hafner.

Fisher, R.A. 1960. *The Design of Experiments,* 8th ed. New York: Hafner.

Frederick, R., J.R. Pagenkopf, and K.L. Perry. 1989. Overview of water quality information systems from a steering committee perspective. Preceedings of the International Symposium on the Design of Water Quality Information Systems, Information Series No. 61, Colorado Water Resources Research Institute, Colorado State University, Fort Collins, Colorado, pages 263–275.

Freund, J.E., 1971. *Mathematical Statistics,* 2nd ed. Englewood Cliffs, New Jersey: Prentice-Hall.

General Accounting Office. 1981. *Better monitoring techniques are needed to assess the quality of rivers and streams.* Report No. CED-81–30, 2 vols., GAO, Washington D.C., April 30.

General Accounting Office. 1986. *The nation's water: key unanswered questions about the quality of rivers and streams.* Report No. GAO/PEMD-86–6, GAO, Washington D.C.

Gilbert, R.O. 1987. *Statistical Methods for Environmental Pollution Monitoring.* New York: Van Nostrand Reinhold.

Grayman, W.M., R.M. Males, R.M. Clark. 1988. Use of microcomputers for the management of water quality data. *Water Resources Bulletin* 24(6):1311–1316.

Green, R.H. 1979. *Sampling Design and Statistical Methods for Environmental Biologists.* New York: Wiley.

Goldman, J.C., T.M. Ray and K.A. Walsh. 1989. Data management for a Federal superfund site: a practical perspective. Proceedings of the International Symposium on the Design of Water Quality Information Systems, Information Series No. 61,

Colorado Water Resources Research Institute, Colorado State University, Fort Collins, Colorado, pages 288–293.

Haney, P.D. and R. Schmidt. 1958. Representative sampling and analytical methods in stream studies. *Sewage and Industrial Wastes,* 30:812–820.

Harcum, J.B. 1990. Water-quality data analysis protocol development. Unpublished PhD thesis, Agricultural and Chemical Engineering Department, Colorado State University, Fort Collins, Colorado.

Harris, J., J.C. Loftis and R.H. Montgomery. 1987. Statistical methods for characterizing ground water quality. *Ground Water,* 25(2):185–193.

Harris, J. 1988. Statistical analysis of ground water quality—interaction of deterministic and stochastic components. Unpublished PhD Thesis, Agricultural and Chemical Engineering Department, Colorado State University, Fort Collins, Colorado, March.

Helsel, D.R. 1987. Advantages of nonparametric procedures for analysis of water quality data. *Hydrological Sciences Journal* 32(2):179–190.

Hines. W.G., D.A. Rickert, and S.W. McKenzie. 1977. Hydrologic analysis and river-quality data programs. *Journal of the Water Pollution Control Federation* 49(9):2031.

Hines, W.W., and D.C. Montgomery. 1980. *Probability and Statistics in Engineering and Management Science,* 2nd ed. New York: Wiley.

Hirsch, R.M., J.R. Slack and R.A. Smith. 1982. Techniques of trend analysis for monthly water quality data. *Water Resources Research,* 18(1):107.

Hirsch, R.M. and J.R. Slack. 1984. A nonparametric trend test for seasonal data with serial dependence. *Water Resources Research,* 29(6):727–732.

Hirsch, R.M. 1988. Statistical methods and sampling design for estimating step trends in surface-water quality. *Water Resources Bulletin* 24(3):493–504.

Hirsch, R.M., W.M. Alley, and W.G. Wilber. 1988. *Concepts for a national water-quality assessment program.* U.S. Geological Survey Circular 1021, available from Books and Open-File Reports Section, USGS, Denver, CO.

Hollander, M., and D.A. Wolfe. 1973. *Nonparametric Statistical Methods.* New York: Wiley.

Howell, D.C. 1982. *Statistical Methods for Psychology,* 2nd ed. Boston: Duxbury Press.

Hurlbert, S.H. 1984. Pseudoreplication and the design of ecological experiments. *Ecological Monographs* 54(2):187–211.

Iman, R.I., and W.J. Conover. 1983. *A Modern Approach to Statistics.* New York: Wiley.

Kerrigan, J.E. (editor). 1970. Proceedings of the National Symposium on Data and Instrumentation for Water Quality Management. Symposium held July 21–23, Madison, Wisconsin, published by the Water Resources Center, Univ. of Wisconsin, Madison, Wisconsin.

Kittrell, F.W. 1969. A practical guide to water quality studies of streams. Report CWR-5, Federal Water Pollution Control Administration, U.S. Department of Interior, Washington, D.C.

Lachance, M., B. Bobeé and J. Haemmerli. 1989. Methodology for the planning and operation of a water quality network with temporal and spatial objectives. Proceedings of the International Symposium on the Design of Water Quality Information Systems, Information Series No. 61, Colorado Water Resources Research Institute, Colorado State University, Fort Collins, Colorado, pages 145–162.

Landwehr, J.M. 1974. Water quality indices—construction and analysis. Unpublished

PhD dissertation, the University of Michigan, Environmental Health Science, 281 pages.

Landwehr, J.M. 1979. A statistical view of a class of water quality indices. *Water Resources Research,* 15(2):460–468.

Larsen, R.J., and M.L. Marx. 1986. *An Introduction to Mathematical Statistics and its Applications,* 2nd ed. Englewood Cliffs, New Jersey: Prentice-Hall.

Lawrence, K.D., T. Holloway, N.H. Crisp, and J. E. Joslin. 1989. The development and investigation into the source and extent of sediment contaminants in an isolated area of Lake Turman, Missouri. *Proceedings of the International Symposium on the Design of Water Quality Information Systems, Information Series No. 61,* Colorado Water Resources Research Institute, Colorado State University, Fort Collins, Colorado, pages 367–381.

Legendre, L., and P. Legendre. 1983. *Numerical Ecology.* Amsterdam: Elsevier.

Lehmann, E.H. 1975. *Nonparametrics: Statistical Methods Based on Ranks,* San Francisco: Holden-Day.

Leo, W.M., R.V. Thomann, and T.W. Gallagher. 1984. Before and after case studies: comparisons of water quality following municipal treatment plant improvements. U.S. Environmental Protection Agency Report No. 430/9–007, Office of Water Program Operations (WH-547), Washington, D.C., May.

Lettenmaier, D.P. 1975. Design of monitoring systems for detection of trends in stream quality. Technical Report 39, Charles W. Harris Laboratory, University of Washington, Seattle, Washington.

Lettenmaier, D.P. 1976. Detection of trends in water quality data from records with dependent observations. *Water Resources Research* 12(5):1037–1046.

Lettenmaier, D.P. 1978. Design considerations for ambient stream quality monitoring. *Water Resources Bulletin* 14(4):884–902. Discussion by D.L. Egar, A.L. Wilson and R.K. Aylesworth, and reply by author in 15(6):1781–1786.

Lettenmaier, D.P., L.L. Conquest, and J.P. Hughes. 1982. Routine streams and rivers water quality trend monitoring review. Technical Report 75. Seattle: Charles W. Harris Laboratory, University of Washington.

Lettenmaier, D.P. 1988. Multivariate nonparametric tests for trend in water quality. *Water Resources Bulletin,* 24(3):505–512.

Loftis, J.C. 1978. Statistical and economic considerations for improving regulatory water quality monitoring networks. Unpublished PhD Thesis, Agricultural and Chemical Engineering Department, Colorado State University, Fort Collins, Colorado.

Loftis, J.C., and R.C. Ward. 1980. Water quality monitoring—some practical sampling frequency considerations. *Environmental Management,* 4(6):521–526.

Loftis, J.C., and R.C. Ward. 1981. Evaluating stream standard violations using a water quality data base. *Water Resources Bulletin* 17(6):1071–1078.

Loftis, J.C., and R.C. Ward. 1982. Routine, fixed-station sampling: an integral part of regulatory water quality monitoring. *Environmental Management* 6(4):279–280.

Loftis, J.C., J. Harris, and R.H. Montgomery. 1987. Detecting changes in ground water quality at regulated facilities. *Ground Water Monitoring Review,* 7(1):72–76.

Loftis, J.C., P.S. Porter and G. Settembre. 1987. Statistical analysis of industrial wastewater monitoring data. *Journal Water Pollution Control Federation,* 59(3): 145–151.

Loftis, J.C., R.C. Ward, R.D. Phillips, and C.H. Taylor. 1989a. An Evaluation of

Trend Detection Techniques For Use in Water Quality Monitoring Programs. U.S. Environmental Protection Agency Report No. EPA/600/3–89/037, Office of Research and Development, Washington, DC, March.

Loftis, J.C., R.D. Phillips, R.C. Ward, and C.H. Taylor. 1989b. WQSTAT II: A water quality statistics package. *Ground Water,* 27(6):866–873.

Loftis, J.C. and C.H. Taylor. 1989. Testing for trend in water quality data. Proceedings of the International Symposium on the Design of Water Quality Information Systems, Information Series No. 61, Colorado Water Resources Research Institute, Colorado State University, Fort Collins, Colorado, pages 461–467.

Lorenz, R., W.R. Sims, T. Killeen and W. Fay. 1989. Managing data from a large-scale groundwater monitoring program. Proceedings of the International Symposium on the Design of Water Quality Information Systems, Information Series No. 61, Colorado Water Resources Research Institute, Colorado State University, Fort Collins, Colorado, pages 309–317.

McBride, G.B. 1989. AQUAL User's and Reference Manual, Version 1.0. Water Quality Centre, Division of Water Sciences, Department of Scientific and Industrial Research, Hamilton, New Zealand.

McBride, G.B., J.C. Rutherford, and R.D. Pridmore. 1988. Modeling organic pollution of streams. In *Civil Engineering Practice: Vol. 5: Water Resources/Environmental.* eds. P.N. Cheremisinoff, N.P. Cheremisinoff, and S.L. Cheng: Lancaster Pa: Technomic Publishing Co.

McGill, R., J.W. Tukey, and W.A. Larsen. 1978. Variations of box plots. *American Statistician,* 32:12–16.

McLeod, A.I., K.W. Hipel and F. Comancho. 1983. Trend assessment of water quality time series. *Water Resources Bulletin,* 19(4):537–547.

McNeely, R.N., V.P. Neimanis and L. Dwyer. 1979. Water Quality Sourcebook—A Guide to Water Quality Parameters. Environment Canada, Inland Waters Directorate, Water Quality Branch, Ottawa, Canada.

Mace, A.E. 1964. *Sample-Size Determination.* New York: Reinhold.

Mar, B.W., R.R. Horner, J.S. Richey, R.N. Palmer, and D.P. Lettenmaier. 1986. Data acquisition: cost-effective methods for obtaining data on water quality. *Environmental Science and Technology* 20(6):545–551.

Martin, G.G., J. Lienhart, T. Hyman and G.C. Levy. 1988. A statistical expert system for data analysis. *American Laboratory,* 20:74–87.

Millard, S.P. 1987. Environmental monitoring, statistics, and the law: room for improvement. *The American Statistician* 41(4):249–253.

Millard, S.P., and D.P. Lettenmaier. 1986. Optimal design of biological sampling programs using the analysis of variance. *Estuarine, Coastal and Shelf Science,* 22:637–656.

Miller, I., and J.E. Freund. 1985. *Probability and Statistics for Engineers,* 3rd ed. Englewood Cliffs, New Jersey: Prentice-Hall.

Miller, D.G., and J.C. Ellis. 1986. Derivation and monitoring of consents. *Water Pollution Control* 85:249–258.

Mills, W.B., G.L. Bowie, T.M. Grieb, K.M. Johnson and R.C. Whittemore. 1986. Handbook—Stream Sampling for Waste Load Allocation Applications. U.S. Environmental Protection Agency Report No. EPA/625/6–86/013, Office of Research and Development, Washington, DC, September.

Montgomery, D.C. 1985. *Introduction to Statistical Quality Control.* New York: Wiley.

Montgomery, H.A.C., and I.C. Hart. 1974. The design of sampling programmes for rivers and effluents. *Water Pollution Control* 33(1):77–101.

Montgomery, R.H. 1987. Uncertainty in ground water quality data. Ph.D. Dissertation. Dept. of Civil Engineering, Colorado State University, Fort Collins, Colorado, 354 pages.

Montgomery, R.H. and J.C. Loftis. 1987. Applicability of the t-test for detecting trends in water quality variables. *Water Resources Bulletin,* 23(4):653, Discussion and Reply in 24(1):201–207.

Montgomery, R.H., J.C. Loftis and J. Harris. 1987. Statistical characteristics of ground water quality variables. *Ground Water,* 25(2):176–184.

Morgan, P.V., B.R. Johnson, H.C. Bramer and W.L. Duncan. 1970. Design of water quality surveillance systems. Water Pollution Control Research Series Report No. 16090 DBJ 08/70, Federal Water Quality Administration, Department of the Interior, Washington, D.C., August.

Morrison, D.E., and R.E. Henkel. 1970. *The Significance Test Controversy.* Chicago Il: Aldin Publishing Company.

Naisbitt, J. 1982. *Megatrends.* New York: Warner Books.

National Academy of Sciences. 1977. Environmental Monitoring: Analytical Studies for the U.S. Environmental Protection Agency, Volume IV. National Academy of Sciences, Washington, D.C.

National Research Council. 1987. *National water quality monitoring and assessment.* National Academy Press, Washington D.C.

Natrella, V.G. 1960. The relation between confidence intervals and tests of significance. *The American Statistician,* 14(1):20–22+.

Neter, J., W. Wasserman, and G.A. Whitmore. 1979. *Applied Statistics.* Boston, Ma: Allyn and Bacon.

Noether, G.E. 1987. Sample size determination for some common nonparametric tests. *Journal of the American Statistical Association,* 82:645.

Noreen, E.W. 1989. *Computer Intensive Methods for Testing Hypotheses.* New York: Wiley.

Norwegian Institute for Water Research. 1987. Manual for Chemical and Biological Monitoring. Prepared by the Programme Centre, Norwegian Institute for Water Research, Oslo, Norway, March.

Oakes, M. 1986. *Statistical Inference: A Commentary for the Social and Behavioural Sciences.* New York: Wiley.

Ott, W.R. 1978. *Environmental Indices: Theory and Practice,* Ann Arbor: Ann Arbor Science Publishers Inc.

Patterson, P., D. Eaker, and D. Esposito. 1989. A groundwater monitoring program for the Santa Cruz River, Pima County, Arizona. Proceedings of the International Symposium on the Design of Water Quality Information Systems, Information Series No. 61, Colorado Water Resources Research Institute, Colorado State University, Fort Collins, Colorado, pages 354–362.

Payne, F.E. and J. Ford. 1988. The concept of TIME—Temporally Integrated Monitoring of Ecosystems—Supplement. Internal report, U.S. Environmental Protection Agency, Environmental Research Laboratory, Corvallis, Oregon.

Pearson, E.S., and H.O. Hartley. 1966. *Biometrika Tables for Statisticians, Volume 1*, 3rd ed. Cambridge, England: Cambridge University Press.

Perry, J.A., R.C. Ward and J.C. Loftis. 1984. Survey of state water quality monitoring programs. *Environmental Management*, 8(1):21–26.

Pielou, E.C. 1984. *The Interpretation of Ecological Data: A Primer on Classification and Ordination*. New York: Wiley.

Plafkin, J.L., M.T. Barbour, K.D. Porter, S.K. Gross and R.M. Hughes. 1989. Rapid bioassessment protocols for use in streams and rivers. U.S. Environmental Protection Agency Report No. EPA/444/4–89–001, Office of Water, Washington, D.C., May.

Pollack, A.K. and J. Ford. 1989. The TIME project—an overview. Proceedings of the International Symposium on the Design of Water Quality Information Systems, Information Series No. 61, Colorado Water Resources Research Institute, Colorado State University, Fort Collins, Colorado, pages 413–424.

Pomeroy, R.D. and G.T. Orlob. 1967. Problems of setting standards and of surveillance for water quality control. Publication No. 36, California State Water Quality Board, Sacramento, California, May.

Ponce, S.L. 1982. The use of paired-basin technique in flow-related wildland water-quality studies. Report WSDG-TP-00004. Fort Collins, CO: USDA Forest Service.

Ponce, S.L. 1980. Water Quality Monitoring Programs. Report WSDG-TP-00002. Fort Collins, CO: USDA Forest Service.

Ponce, S.L. 1982. The Use of the Paired-Basin Technique in Flow-Related Wildland Water-Quality Studies. Report WSDG-TP-00004. Fort Collins, CO: USDA Forest Service.

Porter, P.S. 1986. Statistical analysis of water quality data affected by limits of detection. Unpublished PhD thesis, Department of Agricultural and Chemical Engineering, Colorado State University, Fort Collins, Colorado.

Porter, P.S., R.C. Ward and H.F. Bell. 1988. The detection limit. *Environmental Science and Technology* 22(8):856–861.

Porter, P.S. and R.C. Ward. 1989. Water quality monitoring methods for system error. Proceedings of the International Symposium on the Design of Water Quality Information Systems, Information Series No. 61, Colorado Water Resources Research Institute, Colorado State University, Fort Collins, Colorado, pages 188–211.

Portney, P.R. 1988. Needed: a Bureau of Environmental Statistics. *Resources*, No. 90, Winter, published by Resources for the Future, Washington, D.C., pp. 12–15.

Press, W.H., B.P. Blannery, S.A. Teukolsky and W.T. Vetterling. 1986. *Numerical Recipes*. New York: Cambridge University Press.

Price, B. 1987. Comment on: Environmental monitoring, statistics, and the law: room for improvement. By S.P. Millard. *The American Statistician* 41(4):257–259.

Pridmore, R.D. 1985. Use of the Cochran Q-test in analysing species presence-absence data in freshwater biological studies. In *Biological Monitoring of Freshwaters Part 1*, ed. R.D. Pridmore and A.B. Cooper, Water Quality Centre, Department of Scientific and Industrial Research, Hamilton, New Zealand.

Pridmore, R.D. 1987. Phytoplankton: survey and interpretation. In *Lake Managers Handbook: A Guide to Understanding Investigations into Lake Ecosystems, so as to Assess Management Options for Lakes*. ed. W.N. Vant, Water Quality Centre, Department of Scientific and Industrial Research, Hamilton, New Zealand.

Reckhow, K.H., and S.C. Chapra. 1983. *Engineering Approaches for Lake Management Volume 1: Data Analysis and Empirical Modeling*. Boston, MA: Butterworths.

Reilly, W.K. 1989. Measuring for environmental results. *EPA Journal*, 15(3):2–4.

Rickert, D.A. and W.G. Hines. 1975. A Practical Framework for River-quality Assessment, U.S. Geological Survey Circular 715-A, U.S. Department of Interior, Washington, D.C.

Sachs, L. 1984. *Applied Statistics* (Second Edition). New York: Springer-Verlag.

Sanders, T.G. 1974. Rational design criteria for a river quality monitoring network. Unpublished PhD Thesis, Civil Engineering Department, University of Massachusetts, Amherst, Mass., August.

Sanders, T.G., R.C. Ward, J.C. Loftis, T.D. Steele, D.D. Adrian, and V. Yevjevich. 1983. *Design of Networks for Monitoring Water Quality*. Littleton CO: Water Resources Publications.

Sanders, T.G., R.C. Ward, J.C. Loftis, T.D. Steele, D.D. Adrian, and V. Yevjevich. 1987. *Design of Networks for Monitoring Water Quality* (2nd Edition). Littleton, Colorado: Water Resources Publications.

Scalf, M.R., J.F. McNabb, W.J. Dunlap, R.C. Cosby and J. Fryberger. 1981. Manual of Ground-Water Sampling Procedures. Robert S. Kerr Environmental Research Laboratory, U.S. Environmental Protection Agency, Ada, Oklahoma 74820.

Schilperoort, T. and S. Groot. 1983. Design and optimization of water quality monitoring networks. Publication No. 286, Delft Hydraulics Laboratory, Delft, Netherlands, January.

Schroder, L.J., R.B. See, and T.C. Willoughby. 1989. Identification of the differences between onsite and laboratory pH determinations using a quality assurance program. U.S. Geological Survey Open-File Report 89–409, Reston, Va. Chap. 3.

Sherwani, J.K., and D.H. Moreau. 1975. Strategies for water quality monitoring. Report 107, Water Resources Research Institute, North Carolina State University, Raleigh, North Carolina, June.

Shnider, R.W. and E.S. Shapiro. 1976. Procedures for Evaluating Operations of Water Monitoring Networks. U.S. Environmental Protection Agency, Environmental Monitoring Series No. EPA-600/4–76–050, September.

Smillie, G.M. and M. Flug. 1982. Guidelines for water quality program development in National Park Service areas. WRFSL Report No. 82–2, Water Resources Field Support Laboratory, National Park Service, Colorado State University, Fort Collins, Colorado.

Smith, D.G. 1989. A new form of water quality index for rivers and streams. *Water Science and Technology* 21(2):123–127.

Smith, D.G., G.B. McBride, G.G. Bryers, R.J. Davies-Colley, J.M. Quinn, and W.N. Vant. 1989. Design for a National Water Quality Monitoring Network for New Zealand. Consultancy Report 8016/1, Water Quality Centre, Department of Scientific and Industrial Research, Hamilton, New Zealand.

Smith, D.G. and G.B. McBride. 1989. Initiation of a National Water Quality Monitoring Network for New Zealand. Proceedings of the International Symposium on the Design of Water Quality Information Systems, Information Series. No. 61, Colorado Water Resources Research Institute, Colorado State University, Fort Collins, Colorado, pages 59–72.

Smith, R.A., R.B. Alexander, and M.G. Wolman. 1987. Water-quality trends in the nation's rivers. *Science* 235:1607–1615.

Smith, R.A., R.M. Hirsch and J.R. Slack. 1982. A study of trends in total phosphorus measurements at NASQAN stations. U.S. Geological Survey Water-Supply Paper 2190, U.S. Department of the Interior, Washington, D.C.

Snedecor, G.W., and W.G. Cochran. 1980. *Statistical Methods*. 7th ed. Ames, Iowa: The Iowa State University Press.

Sokal, R.R., and F.J. Rohlf. 1969. *Biometry*. San Francisco, Ca: W.H. Freeman and Co.

Spiegel, M.R. 1961. *Schaum's Outline of Theory and Problems in Statistics*. New York: McGraw Hill.

Stednick, J.D. and L.C. Roig. 1989. Kalman filter calculation of sampling frequency when determining annual mean solute concentrations. *Water Resources Bulletin*, 25(3):675–683.

Steel, R.G.D., and J.H. Torrie. 1960. *Principles and Procedures of Statistics*. New York: McGraw Hill.

Stewart-Oaten, A., W.W. Murdoch, and K.R. Parker. 1986. Environmental impact assessment: "pseudoreplication" in time? *Ecology* 67(4):929–940.

Taylor, P.L. 1970. Experience with a water pollution control storage and retrieval system. Proceedings of the National Symposium on Data and Instrumentation for Water Quality Management, Water Resources Center, University of Wisconsin, Madison, Wisconsin, December, pages 198–207.

Taylor, J.K. and T.W. Stanley (editors). 1985. *Quality Assurance for Environmental Measurements*. ASTM Special Technical Publication 867, American Society for Testing and Materials, Philadelphia, Pennsylvania.

Thompson, S.M. and M.W. Rodgers. 1985. Micro TIDEDA user's manual. Hydrology Centre, Department of Scientific and Industrial Research, Christchurch, New Zealand.

Thornton, K.W., F.E. Payne, J. Ford, and D.H. Landers. 1987. The concept of TIME (Temporal Integrated Monitoring of Ecosystems). Aquatic Effects Research Program, U.S. Environmental Protection Agency, Corvallis Environmental Research Laboratory, Corvallis, Oregon, August.

Toft, C.A., and P.J. Shea. 1983. Detecting community-wide patterns: estimating power strengthens statistical inference. *The American Naturalist* 122(5):618–625.

Tracey, R., and B. McCammon. 1990. The Bull Run watershed as a water quality information system. Proceedings of the International Symposium on the Design of Water Quality Information Systems, Information Series No. 61, Colorado Water Resources Research Institute, Colorado State University, Fort Collins, Colorado, pages 348–353.

Tufte, E.R. 1983. *The Visual Display of Quantitative Information*. Cheshire, Ct: Graphics Press.

Tukey, J.W. 1977. *Exploratory Data Analysis*. Reading, Mass: Addison-Wesley.

U.S. Department of Energy. 1989. Assessment of Environmental Conditions at the Rocky Flats Plant. U.S. Department of Energy, Special Assignment Environmental Team, Golden, Colorado, August.

U.S. Environmental Protection Agency. 1974. Digest of EPA's Monitoring-related Statutory Authority. U.S. Environmental Protection Agency, Office of Research and Development, Office of Monitoring Systems, Washington, D.C.

U.S. Environmental Protection Agency. 1975. Model State Water Monitoring Program. U.S. Environmental Protection Agency Report No. EPA-440/9–74–002, Washington, D.C.

U.S. Environmental Protection Agency. 1977. Basic Water Monitoring Program. U.S. Environmental Protection Agency Report No. EPA 440/9–76–025, Washington, D.C.

U.S. Environmental Protection Agency. 1979a. Methods for Chemical Analysis of Water and Wastes. Environmental Monitoring and Support Laboratory, Cincinnati, Ohio, Report No. EPA-600/4–79–020, March.

U.S. Environmental Protection Agency. 1979b. NPDES compliance sampling inspection manual. Report No. MCD-51, Enforcement Division, Office of Water Enforcement, Compliance Branch, Washington, D.C.

U.S. Environmental Protection Agency. 1980. Oil Shale Symposium: Sampling, Analysis, and Quality Assurance. Industrial Environmental Research Laboratory, Cincinnati, Ohio, Report No. EPA-600/9–80–022, June.

U.S. Environmental Protection Agency. 1981. Methods for organic chemical analysis of water and wastes by GC, HPLC and GC/MS. Environmental Monitoring and Support Laboratory, Cincinnati, Ohio.

U.S. Environmental Protection Agency. 1982. Handbook for Sampling and Sample Preservation of Water and Wastewater. Report No. EPA-600/4–82–029, Environmental Monitoring and Support Laboratory, Cincinnati, Ohio, September.

U.S. Environmental Protection Agency. 1983. Technical support manual: waterbody surveys and assessments for conducting use attainability analyses. Office of Water Regulations and Standards, Washington, D.C., November.

U.S. Environmental Protection Agency. 1985a. Guidance: 1986 state water quality assessments (Section 305(b) Reports). Office of Water Regulations and Standards (WH-553), Washington, D.C., June.

U.S. Environmental Protection Agency. 1985b. Guidance for State Water Monitoring and Wasteload Allocation Programs. U.S. Environmental Protection Agency Report No. EPA 440/4–85–031, Office of Water Regulations and Standards, Washington, D.C., October.

U.S. Environmental Protection Agency. 1986a. RCRA ground-water monitoring technical enforcement guidance document. Office of Solid Waste and Emergency Response Report 9950.1, Washington, D.C., September.

U.S. Environmental Protection Agency. 1986b. Users' manual: monitoring checklist database program. Office of Water Regulations and Standards, Washington, D.C., December.

U.S. Environmental Protection Agency. 1987a. Surface Water Quality Monitoring: a Framework for Change. Office of Water Enforcement and Permits, Office of Water Regulations and Standards, USEPA, WH-553, Washington, D.C., September.

U.S. Environmental Protection Agency. 1987b. Handbook of Methods for Acid Deposition Studies—Laboratory Analysis for Surface Water Chemistry. U.S. Environmental Protection Agency Report No. EPA 600/4–87/026, Office of Research and Development, Washington, D.C., September.

U.S. Environmental Protection Agency. 1989. Statistical analysis of ground-water monitoring data at RCRA facilities. Interim Final Guidance, Office of Solid Waste, Waste Management Division, U.S. EPA, Washington, D.C., 20460, February.

U.S. Geological Survey. 1977. National Handbook of Recommended Methods for Water-Data Acquisition. Office of Water Data Coordination, Geological Survey, U.S. Department of the Interior, Reston, Virginia (as updated).

van Belle, G., and J.P. Hughes. 1983. Monitoring for water quality: fixed stations versus intensive surveys. *Journal Water Pollution Control Federation* 55(4):400–404.

van Belle, G., and J. P. Hughes. 1984. Nonparametric tests for trend in water quality. *Water Resources Research* 20:127–136.

Velleman, P.F., and D.C. Hoaglin. 1981. *Applications, Basics, and Computing of Exploratory Data Analysis.* Boston, Ma: Duxbury.

Velz, C.J. 1950. Sampling for effective evaluation of stream pollution. *Sewage and Industrial Wastes,* 22:666–682.

Virginia Water Control Board. 1988. Water quality assessment operating procedures manual. Richmond, Virginia, August.

Wallis, W.A., and H.V. Roberts. 1956. *Statistics: A New Approach.* New York: The Free Press.

Ward, R.C. 1973. Data acquisition systems in water quality management. Socioeconomic Environmental Studies Series Report No. EPA-R5–73–014, U.S. Environmental Protection Agency, Washington, D.C., May.

Ward, R.C. 1979. Regulatory water quality monitoring: a system's perspective. *Water Resources Bulletin* 15(2):369–380.

Ward, R.C., G.M. Smillie, J.C. Loftis and T.G. Sanders. 1982. Relating Stream Standards to Water Quality Monitoring Practices. National Science Foundation Final Report for Grant Number PRA-7913073, January.

Ward, R.C. and J.C. Loftis. 1983. Incorporating the stochastic nature of water quality into management. *Journal Water Pollution Control Federation,* 55(4):408–414.

Ward, R.C. 1986. Framework for designing water quality information systems. Proceedings of the 2nd Scientific Assembly entitled "Monitoring to detect changes in water quality time series" Publication No. 157, International Association of Hydrological Sciences, Wallingford, Oxfordshire, UK, pages 89–98.

Ward, R.C. and G.B. McBride. 1986. Design of water quality monitoring systems in New Zealand. Publication No. 8, Water Quality Centre, Ministry of Works and Development, Hamilton, New Zealand, April.

Ward, R.C. and J.C. Loftis. 1986. Establishing statistical design criteria for water quality monitoring systems: review and synthesis. *Water Resources Bulletin,* 22(5):759–769.

Ward, R.C., J.C. Loftis, and G.B. McBride. 1986. The "data-rich but information-poor" syndrome in water quality monitoring. *Environmental Management* 10(3):291–297.

Ward, R.C., J.C. Loftis, H.P. DeLong and H.F. Bell. 1988. Groundwater quality: a data analysis protocol. *Journal Water Pollution Control Federation,* 60(11):1938–1945.

Ward, R.C. 1989. Future Directions for Water Quality Monitoring in Virginia. Final Report for Contract No. 414-S-90–002, Virginia Water Control Board, Richmond, Virginia, July.

Ward, R.C. and J.C. Loftis. 1989. Monitoring systems for water quality. CRC *Critical Reviews in Environmental Control,* 19(2):101–118.

Ward, R.C., J.C. Loftis and G.B. McBride (Editors). 1989. Proceedings of the International Symposium on the Design of Water Quality Information Systems. Information Series No. 61, Colorado Water Resources Research Institute, Colorado State University, Fort Collins, Colorado.

Warn, A.E. and P.J. Matthews. 1984. Calculation of the compliance of discharges with emission standards. *Water Science and Technology,* 16:183–196.

Whitfield, P.H. 1988. Goals and data collection designs for water quality monitoring. *Water Resources Bulletin,* 24(4):775–780.

Welkowitz, J., R.B. Ewen, and J. Cohen. 1982. *Introductory Statistics for the Behavioral Sciences,* 3rd ed. New York: Academic Press.

Wilson, L.G. 1980. Monitoring in the Vadose Zone. U.S. Environmental Protection Agency Report No. EPA-600/7–80–134, Environmental Monitoring Systems Laboratory, Las Vegas, Nevada, June.

Wolman, M.G. 1971. The nation's rivers. *Science,* 174:905–918

World Health Organization. 1978. Water quality surveys. Studies and Reports in Hydrology No. 23, World Health Organization, 1211 Geneva 27, Switzerland.

Zar, J.H. 1984. *Biostatistical Analysis,* 2nd ed. Englewood Cliffs, New Jersey: Prentice-Hall.

Appendix A
SAMPLE SIZE CALCULATION FORMULAE

A.1 TESTING NOMINAL SCALE DATA

Consider testing compliance with a standard that requires a water quality variable to be within a specified limit or range for at least a certain proportion of some compliance assessment period (as used in England and Wales, see Miller and Ellis 1986). Call this proportion p^*. Compliance is to be inferred if there are no more than certain number (e) of exceedances of the limit or allowable range in a batch of n random samples. The compliance rule is to be designed given the error risks α and β, as follows.

- If the effluent was "borderline," i.e., the water quality variable was within the limit or range exactly $100p^*\%$ of the time, the risk of falsely inferring that the standard had been breached is to be 100α. Such a false inference would occur if, by chance, more than e exceedances were found in the batch when the effluent was really in compliance.
- If the standard has been breached; i.e., the variable was within the limit or range for only $100\,p^*\%$ of the period, where $p < p^*$, the risk of falsely inferring compliance is to be $100\beta\%$. Such a false inference would occur if, by chance, no more than e exceedances were found in the batch when the effluent was really out of compliance.

To calculate the required n, first calculate e for a range of n, by taking the largest e satisfying

$$1 - B(e;n,\, 1 - p^*) \leq \alpha \tag{A-1}$$

where B is the cumulative binomial probability. Tables for this probability can be found in standard texts (such as Miller and Freund, 1985) but usually only for n up to 20, and p in increments of 0.05. It may therefore be best to use an algorithm to obtain B for any values of e, n and p^* (Press et al. 1986).

Then to calculate the appropriate value of the sample size select the particular value of n that most nearly satisfies

$$\beta = B(e;n,\, 1 - p) \tag{A-2}$$

The results of the calculation using Eq. A-2 can be displayed graphically, enabling the appropriate e and n values to be read for given values of α, β, p^* and p (Miller and Ellis, 1986).

It may be noted that the sample compliance ratio $1 - e/n$ approaches p^* from below as n approaches infinity. For 16 samples measuring compliance with a 95th percentile standard, the ratio is only 87.5%.

A.2 ESTIMATION AND TESTING—ORDINAL AND INTERVAL SCALE DATA

Exact sample size calculation formulae are generally available only for parametric methods. The formulae below, therefore, apply in a strict sense to means only. However, for sample size calculations where nonparametric methods will be used, one might use the parametric formulae as a rough approximation. Alternatively, one might use the approximate sample size formulae for nonparametric methods given by Lehmann (1975) and Noether (1987).

One of the reasons that nonparametric tests are so appealing is that water quality populations are often right-skewed. Thus, while the sample estimate of variance is unbiased, one can expect a majority of the estimates to underestimate the population value by a small amount. Occasionally, though, the estimate will be much higher. This could mean that most calculated sample sizes will be too low. Bros and Cowell (1987) have suggested using randomizations of pilot survey data to get optimal size.

The formulae usually are implicit and have to be solved iteratively—they have n (the sample size) on both sides. Three iterations usually suffice. The answer will seldom be an integer, and so it should be rounded up. If $n > 30$, the formulae can be made explicit, by replacing the t distribution percentile by the normal distribution percentile, and so can be solved directly.

A.2.1 Estimation

The question is how many samples are needed to obtain a sufficiently precise estimate of the true mean, or difference in means. The confidence interval (CI) half-width is the usual measure of precision, and the sample size formulae are obtained directly from the CI formulae. In all cases there are two data requirements:

- w/s = ratio of CI half-width to a sample standard deviation;
- α = probability that the CI does not contain the population mean.

If s is known (through a pilot study, or by extrapolation from another study) one can select n to obtain the desired value of w. Otherwise, one chooses a desired value of the ratio w/s. A value on the order of unity is typical.

To estimate the mean of a normally distributed random variable from a single data set (single sample):

$$n = s^2[t_{\alpha/2,(n-1)}]^2/w^2 \qquad \text{(A-3)}$$

(Mace 1964:36), where s is the standard deviation of the sample to be collected.

To estimate the difference between two means of normal populations with equal variances, from two samples:

$$n = 2s_p^2[t_{\alpha/2,2(n-1)}]^2/w^2 \qquad \text{(A-4)}$$

where n is the size of *each* sample (so the total number of samples is $2n$), w is the half-width of the CI about the difference to be estimated, and s_p^2 is the pooled variance of the two samples to be collected. Note that this variance does not include the variability caused by the difference—it merely takes the variance of the data in each sample about its mean, and averages (i.e., pools) the two. If the two sample sizes (n_1 and n_2) are constrained to be unequal, then use the following expression, derived from the harmonic mean sample size (Zar 1984:133):

$$n_2 = nn_1/(2n_1 - n) \qquad \text{(A-5)}$$

where n_1 must be greater than $n/2$ for this formula to work.

Note that in both equations 3 and 4, n increases as s increases but decreases with w or α.

To illustrate, consider the following example. Calculate the sample size necessary to estimate the difference between two means with a precision ratio $w/s_p = 1$, and $\alpha = 5\%$. Required values of the t statistic can be found in Snedecor and Cochran (1980:Table A-4). (The referenced table is two-tailed, thus the probability level corresponding to the problem is 0.05. To use a one-tailed table, the required probability level is 0.025.)

For $n = 20$, then Eq. A-4 gives $n = 2[t_{0.05/2,38}]^2 = 2[2.024]^2 = 8.19$. So a better guess is $n = 9$. Then Eq. A-4 gives $n = 2[2.120]^2 = 8.98$. So the answer is $n = 9$. If n_1 is constrained to be 7, then from Eq. A-5 one gets $n_2 = 12.6$, which is rounded up to 13. So with unequal sample sizes, one gets 20 total samples, instead of 18 for even sample sizes. The total sample size is always smallest if each sample is of the same size.

A.2.2 Testing

The question is how many samples does one need to be able to detect a certain difference, with stated error risks? The following sample size formulae are derived from approximations to the relevant OC curves. In all cases there are three data requirements:

- $d = \Delta/s$ = relative difference to be detected, where delta is the true difference and and s is the sample standard deviation of the residuals (the data with the difference removed);
- α = Type I error risk of inferring a difference when none exists;
- β = Type II error risk of failing to detect a difference when in fact it equals d.

To test the difference between a normal mean and a standard, from a single sample:

$$n = s^2[t_{\alpha/2,(n-1)} + t_{\beta,(n-1)}]^2/\Delta^2 \qquad \text{(A-6)}$$

(Zar 1984:110). The sample size increases with increasing s, and decreases with increasing α, β, and Δ. Note that if the Type II error risk is not specified then the power to detect d takes the default value of 50% (because for the second t term in Eq. A-6 to be zero, β must be 0.5). In other words, if the reference value lies exactly at the edge of the confidence interval, it will have only a 50% chance of being found in that interval. This formula also holds for testing the difference between paired data, but with s being the standard deviation of the differences between the pairs.

To test the difference between two normal means from populations with equal variances:

$$n = 2s_p^2[t_{\alpha/2,2(n-1)} + t_{\beta,2(n-1)}]^2/\Delta^2 \tag{A-7}$$

(Zar 1984:134), where n is the size of each sample, and, as in Eq. A-5, s_p^2 is the pooled variance of the two samples to be collected. If the difference is spatial, one takes n samples at each site for a combined sample size of $2n$. If the difference is temporal, i.e., between two time periods at the same site—the periods being before and after some step change that may have occurred—the total sample size is still $2n$. Again, the smallest total sample size occurs if each sample is of the same size; if not, Eq. A-5 can be used. Also, if power is not specified, it defaults to 50%.

To test for a linear trend in a variable distributed normally about the trend line:

$$n = 12s^2[t_{\alpha/2,(n-2)} + t_{\beta,(n-2)}]^2/\Delta^2 \tag{A-8}$$

where s is the standard deviation of the time series in the absence of the trend, and Δ is the magnitude of the trend over the whole period of record (as may be derived from Lettenmaier 1976).

Note that in these formulae also, n increases as s increases, but decreases with increasing α, β, or Δ. Also, the total number of samples needed for detecting a linear change is about three times that needed for detecting a step change—s_p/Δ in Eq. A-7 equal to s/Δ in Eq. A-8 and ignoring the difference in the degrees of freedom.

To again illustrate with an example, consider the following. Take $\Delta/s = 1$, $\alpha = \beta = 10\%$. Guess $n = 80$. Then Eq. A-8 gives $n = 12(1.665 + 1.292)^2 = 104.9$. So try $n = 105$. Then Eq. A-8 gives $n = 12(1.659 + 1.290)^2 = 104.4$. So $n = 105$. Note that had normal percentiles been used instead of t, the slightly less accurate result would be obtained: $n = 12(1.645 + 1.282)^2 = 102.8$, say $n = 103$.

A.3 NEED FOR INDEPENDENT SAMPLES

All of the foregoing sample size calculations assume independent samples. That is to say that the observations must be strictly random or spaced far enough apart that no information is carried over from sample to sample. If samples are serially correlated, as discussed in Chapter 6, then the "effective number of independent samples" (Bayley and Hammersley, 1946) will be smaller than the number of samples collected. An approach to sampling frequency selection for trend detection is that of Lettenmaier (1976, 1978), who assumed that the water quality variable time series follows a simple Markov process. The effect of serial correlation on sample sizes needed to estimate long-term means is discussed in Loftis and Ward (1980) and Sanders et al. (1983).

Appendix B
COMPUTERIZING WATER QUALITY DATA MANAGEMENT AND ANALYSIS

Given the availability of powerful and inexpensive hardware and software, it is almost unthinkable to design a water quality information system without making use of computers, especially microcomputers or workstations, for data management and analysis.

B.1 COMMERCIALLY AVAILABLE SOFTWARE

B.1.1 Spreadsheets

For starters, spreadsheets such as Lotus 1–2–3 or Borland's Quattro can be used to store and retrieve data, produce time series plots, and perform simple statistical analyses. Some basic summary statistics (means, standard deviations) and graphical analyses (histograms) are very straightforward, and other simple analyses can be programmed using microcomputers.

B.1.2 Relational Data Bases

At a somewhat higher level, relational data bases, such as DBase III or IV or Rbase for DOS, provide more powerful data management capabilities. These programs may be used "as is" in an interactive mode for simple storage and retrieval, or programs may be written to automate data entry, analysis, and generation of reports. (See for example, Loftis, et al., 1987.)

A disadvantage of such software is that routines for statistical analyses have to be written from scratch. Furthermore, graphical capabilities may be limited. Production of software for effective storage and retrieval of water quality data is a difficult task. The authors present some of the most important considerations later under "water quality archive design criteria."

B.1.3 Statistical Software Packages

Many options now exist for purchasing statistical software. Powerful packages for mainframes, notably SAS and SPSS, have been released in microcomputer and/or workstation versions and are particularly well suited for users with mainframe experience on the same software. For Fortran programmers, IMSL provides a large array of subroutines for mathematical and statistical analysis. For the less experienced or less ambitious user, integrated packages, such as Statgraphics, provide a complete range of statistical methods with built-in graphics and a menu-driven interface requiring no commands.

Statistical software packages, while providing most of the desired statistical methods, typically have limited capabilities for data management. It may therefore be desirable in some instances to use both a statistics package and a relational data base to obtain a complete data management system.

B.1.4 Specialized Software for Water Quality

Most of the higher quality statistical software packages now include nonparametric methods and are well suited to water quality data analysis. However, with any of these packages there are disadvantages such as high initial cost, failure to include some methods which are very useful for water quality, and significant time commitment required to learn the system.

Therefore statistical software is now being developed specifically for water quality applications, incorporating those methods found to be most useful in this specialized area. An early example of water quality software is Fortran code presented by Lettenmaier et al. (1982) for water quality trend analysis. A more recent package of data management and statistical programs is described by Alexander et al. (1989). The package is oriented toward trend analysis in both censored and uncensored water quality data.

B.2 WQStat II

A package called WQStat II, developed at Colorado State University, provides an inexpensive and easy method of performing routine water quality data summaries and analysis on microcomputers. WQStat II is a self-contained data management and data analysis system designed for water quality management professionals with a moderate level of statistical training. The program is menu-driven and requires no commands.

WQStat II is not an expert system, per se. However, the software does incorporate a high level of statistical expertise by including only those data analysis techniques which are most appropriate for water quality management and organizing them according to the three information objectives discussed throughout this book—describing average conditions, evaluating changes in water quality, and evaluating extreme conditions.

The program additionally provides information to the user who may be unfamiliar with some of the statistical procedures via "Consult the Expert" options in many of the menus. In some cases the "Expert" uses simple rules to guide the user in choosing from available statistical routines and in interpreting their results.

WQStat II relies heavily upon graphical and nonparametric procedures to obtain the greatest possible applicability to water quality data.

B.2.1 Data Management

In WQStat II water quality observations for a single variable and location are associated with a single user-defined file name. The program provides a template for manual data entry and editing. Data can also be imported from ASCII files or Lotus 1–2–3 spreadsheet files.

Observations can be entered at any frequency from daily to quarterly and need not be evenly spaced in time. However, WQStat II will prepare data for analysis by creating either a monthly or quarterly record according to the user's specifications. "Quarters" are user-defined to correspond to seasons and need not be three months long.

When preparing data for analysis, WQStat II presents a time series plot of the monthly or quarterly series. The user can select any desired segment(s) of the series for analysis using arrow keys to move an indicator pointer on the screen.

B.2.2 Summary Statistics

WQStat II provides the following summary statistics options for investigating the general distribution of a water quality variable.

1. Mean, standard deviation, skewness and kurtosis with tests for normality;
2. time series plot;
3. seasonal box-and-whisker plot (indication of seasonality);
4. annual box-and-whisker plot (indication of trends);
5. histogram;
6. correlogram or sample autocorrelation function (indication of serial correlation); and
7. Kruskall-Wallis test for seasonality.

B.2.3 Trend Analysis

WQStat II offers two options for trend analysis on a selected data segment:

1. Kendall-tau and seasonal Kendall tests;
2. Analysis of covariance on ranks of data.

Both tests are performed at the 95%, 90%, and 80% confidence levels (5%, 10%, and 20% significance levels, respectively).

B.2.4 Median Analysis

The following options provide comparisons of average conditions across two or more data segments.

1. Wilcoxon signed rank test for comparing two data segments which can be paired;
2. Mann-Whitney test for comparing two data segments which cannot be paired; and
3. Kruskall-Wallis test for comparing 3–20 data segments simultaneously.

Comparisons can be made across time or space; i.e., within a single WQStat II data file, between two files, or among files.

B.2.5 Excursion Analysis

Two options are provided for evaluating the significance of extreme values. First the program can compute the fraction or proportion of observations in each season which exceed some specified action level. Confidence limits (99% and 95%) are then given for those proportions.

The second option computes tolerance intervals, by season, for a specified segment. A tolerance interval on the difference between two files (locations) may be used to check for impacts on one location (a downgradient well) with respect to another (an upgradient well). The user may select either parametric or nonparametric tolerance intervals. This tolerance interval approach to evaluating changes in ground water quality at regulated facilities is described in Loftis, et al. (1987).

B.2.6 Availability and Terms of Use

At the time of this writing, the software and user's manual may be purchased from Colorado State University. Please write to the authors for current information on availability and pricing—Room 100 Engineering South, Colorado State University, Fort Collins, Colorado 80523.

Users are free to copy and distribute WQStat II if no fee is charged for use, copying, or distributing and WQStat II is not modified in any way.

Colorado State University makes no warranties, expressed or implied, including, but not limited to, merchantability or fitness for any particular purpose. In no event shall Colorado State University or its employees be liable for indirect or consequential damages arising from the use of the software program.

Furthermore, no user support is provided. The software may be modified in the near future, and new versions may differ from WQStat II in both features and cost. Colorado State University may also choose to stop distributing WQStat II and/or future versions at any time.

Since WQStat II was developed on a small budget as "share-ware," it is limited in capabilities for data analysis and report generation compared to commercial packages such as Statgraphics. Furthermore, the presence of "bugs" in the code results in occasional system crashes and dictates that results must be checked by hand or other software for important projects.

In spite of these limitations, a few hundred users are now enjoying the benefits of a package developed specifically for water quality, and most of the feedback from users has been very positive. In the future, we will probably see the development of commercial software designed specifically for water quality applications, thus affording a higher level of performance and support than that currently available.

B.3 WATER QUALITY DATA ARCHIVE DESIGN CRITERIA

In the past, once water quality data had been reported from a laboratory or a field party, they had been recorded in reports and on paper files and stored in cabinets, store-rooms, and on shelves. But as water quality monitoring has assumed a more important place in water resources management, there has been a realization that more permanent and accessible storage is required. Without that, the many analyses that data may be subjected to—some for purposes not envisaged when the data were collected—cannot be performed. Worse still, as time goes by the existence of the data may not be known, and even if it is obtainable, it may be so poorly documented as to be of no use.

Some years ago this realization prompted some large organizations to develop water quality data archive software for their mainframe computers (e.g., STORET by the U.S. EPA, WATSTORE by the U.S. Geological Survey, and NAQUADAT by Environment Canada).

More recently, the widespread availability of cheap powerful computing has promoted many smaller organizations to develop computer based archives on smaller machines, using commercially available database languages (Goldman et al, 1989; Lorenz et al, 1989; and Patterson et al, 1989). Some older mainframe systems are being revised (e.g., Frederick et al, 1989 and Brooksbank, 1989), but these are not to be available on smaller machines.

In New Zealand, some considerable effort has been put into the design and construction of water quality data archive software, called AQUAL (McBride 1989). This has involved substantial discussion with prospective users. The hope is that it will eventually arrive at software that can be run on personal computers (at present it runs on VAX machines only), so that it becomes the preferred method of data storage by many organizations. Experience in New Zealand with an archive for hydrological data (TIDEDA—Thompson and Rodgers, 1985) has shown the enormous advantages for data management, such as widespread use of that product by many organizations, improved documentation and interpretation of data, greater awareness of just how much information *can* be extracted from data, and improved data security. It is felt that that provision of a good water quality data archive software would have a similarly beneficial effect. Of course, there would always be a need for organizations to build special features for their needs, but if that can be "bolted on" to common archive software, effort can be spent where it is needed.

The efforts to date on building such software, at least for smaller machines, do not indicate that the important issues have all been addressed. In the New Zealand effort to build a water quality data management system, the following features were identified as being critical to its future success.

There are six over-riding requirements, of which the first is paramount. These are the following:

1. *Data must be stored and retrieved unambiguously.* A vast number of water quality variables need to be stored. For each of them, different preparation methods (e.g., filtering) and analysis methods (e.g., using different reagents) may be used, there may be many choices of units in which they can be reported, results may be reported below or above a detection limit, and it is necessary to clearly identify where, when and how the sample was taken. Unless there is full documentation of these items, the data can quickly become ambiguous, and their usefulness compromised. The main features to be incorporated to achieve this are:
   - distinguish mandatory and optional input. The former is data that the system must have in order to store and retrieve data unambiguously. It means that the user will be prohibited from updating the database, or from submitting an enquiry,

until all necessary data have been supplied. The latter is information that may be useful in interpreting the data; e.g., comments on unusual conditions, expected range of data;

- for censored data the $<$ and $>$ signs should be stored with the detection limit, and never separated from it. Ideally, the system should allow storage of alphanumeric sample values, so that nominal and ordinal scale data can be stored without having to use numerical codes;
- the system must ensure that the data stored and retrieved are documented so that there can be no confusion as to what they are (e.g., merely storing total phosphorus data with the nickname "TP" is ambiguous; laboratory methods for TP have changed with time, and the method used should be explicitly documented).

2. *The software must be portable.* The multiplicity of computer operating systems, languages and dialects is a challenge that will have to be addressed, to save substantial reprogramming efforts for different installations. New database languages, such as ORACLE (as the redesign of NAQUADAT is using—Brooksbank 1989) may offer considerable advantages here, being portable across operating systems. But the special features of water quality data—missing data, censored data ("less than" and "greater than"), and alphabetical values (some nominal and ordinal scale data) must be able to be handled in such a language. Further, the code should not have any local conditions built in, so an important feature is:
   - all data concerning containing lists of codes (e.g., for agencies, regions, attributes, site types), allowable units and units conversion factors should be stored in an *external* file. These data must not be built into the code directly, because this necessitates reprogramming to use local values. Instead, the external file should be able to be modified and updated as necessary.
3. *The software must be easy-to-use.* There is no better way to discourage prospective users than to force them to learn complicated procedures to even get the system to run, and then have to go through laborious and repetitive procedures to enter and retrieve data. The system must be user-friendly. There is of course the potential for conflict between this ease-of-use requirement and that of sorting data unambiguously: the system must demand the information it needs to ensure that its database does not receive ambiguous data. The requirement to give it such data cannot be avoided, but any "inconvenience" a user might see in this can be minimized by adopting these ease-of-use features:
   - routine data entry must not be discouraged, but to ensure that data already archived that have in some manner been certified are not corrupted by new poor quality data, a separate verification step should be provided. Routine data should automatically be flagged as "unverified" when first entered;
   - in entering data for sites, samples and projects, identifiers can be used, preceded by an agency code so the user knows whose data these are. In entering and retrieving data for water quality variables, nicknames should be used. However, data should *not* be stored by nickname, because that can promote ambiguity as analysis methods change; rather they should be stored by a unique code that is indexed to a full description of the method. Because it is desirable from time to time to detach a nickname from an old code and reattach it to a new one, considerable care is needed to ensure that there is always a unique attachment of nicknames to codes;

- menus of options, so that users can quickly see what options are available, and how to get them;
- full screen data entry into fields, where default values are displayed and overwritten only if necessary;
- provision of on-line help, by way of: (1) a help item in the menu giving general information on the system, and (2) special keys to speed up entry of data onto a data entry screen. This includes keys to: *toggle* through options (e.g., units of ppm, ppb), *repeat* the last entry (especially useful for entry of dates), and *increment* the last entry (especially useful when entering sequential data on dates and sample numbers);
- use of prompts—to warn a user when something inappropriate has been attempted, to post information that may be useful to the user (e.g., displaying the full name of a variable when its nickname is entered), and to advise successful completion of database updates and enquiries;
- minimizing response time and storage requirements of the system. This means using a relational database management configuration using indexed files for the four main categories of data: sites, water quality variables, projects, and samples;
- sample data entry is simplest if the sample is the organizing concept for the database (Grayman, Males, and Clark 1988): in the usual case of more than one analysis being performed on a sample, other mandatory information about the sample (e.g., date and place of sampling) does not have to be repeated. Once the sample is "registered" that information is known and need not be repeated each time values for the variables are entered;
- error logging—if an error is made when the database is updated, that fact and some diagnostic information is stored in an error log file that may then be browsed;
- error trapping—when an error occurs the user is not thrown off the system;
- a fully-documented users and reference guide—this should not need to be referred to often, because the system should aim to be self-explanatory. But it should give examples using a demonstration dataset that is supplied with the code, and give full documentation of procedures used.

4. *Protection against wilful or accidental damage.* By adopting the convention that the routine entry of sample data is open to all users, but ensuring that new data is flagged as "unverified," one can guard against corrupting existing "verified" data. So the process of "verification" should not be open to any user: it should be a privileged operation (e.g., by special password). The criteria used to verify the data should not be dictated by the system. There are other operations that should be privileged also. These are:
   - entry of bulk data—the database can be flooded by the sudden inexpert entry of a large mass of data;
   - editing of sample data—this should only be possible on unverified data, and it may need to be protected;
   - registration of sites—where a full description of the site is made, including map references and the site type (river, effluent, . . .). If a code can be devised to indicate the connectivity of sites (e.g., to show whether one site is upstream or downstream of another, or whether they are on different waterways), that can greatly facilitate subsequent enquiries;

- registration of the water quality variables—the values of the variables to be entered will need to be stored with a unique code and consistent units. So in first registering the variable a code will have to be defined for that variable. The code defines a particular method and sample preparation: if either change, a new code would have to be registered. Sample data do not have to be entered in the registered units, but the system must convert them to the registered units in which data are to be stored. The code must be defined in a separate document. This registration should also give a full name to the method (for use in prompts and reports) and attach a nickname to the code;
- registration of projects—to assign water quality variables to a project. This facilitates subsequent project reports.

5. *Unambiguous outward data transfer*. The archive need not be viewed as a substantial data analysis tool—any non-standard analysis can be performed by transferring the data outward to a data analysis package. But in so doing, one must be aware that such packages usually have no features for handling censored data. Sometimes even records with missing data are not handled well. Options could be provided to set the "less than" data to half their detection limit (a common practice), or, perhaps better, to make two output data sets: one with all "less than" data set to zero, and another with them all set to the detection limit. Comparing analyses on both sets enables the effect of the censoring to be gauged. Whatever is done must be recorded in the outward dataset. A warning ought to be posted on the screen and in the output file if "greater than" data are present.
6. *Flexible enquiry and reporting*. There should be flexible procedures for selecting data. It should be possible to obtain a list of sites or samples meeting logical criteria that the user may specify. Thus if a river connectivity code has been defined in the site registration, a search could be made for all DO < 5 ppm on a particular river since, say, 1 Jan. 1983. Provision should also be made for a user-defined report, where summary statistics can be reported. Simple correlation and regression statistics would be of benefit, especially if there is a procedure for performing these on the ranks of the data (so that nonparametric regression and correlation is performed). Simple variable versus time, and variable versus variable plots should be possible.

Hopefully these suggestions will be of help to readers who are faced with the task of creating an effective water quality data management system.

Disclaimer: The use of trade names in our discussion of software does not imply an endorsement by the authors or publisher.

INDEX